FOOD SCIENCE AND TECHNOLOGY

LYCOPENE

FOOD SOURCES, POTENTIAL ROLE IN HUMAN HEALTH AND ANTIOXIDANT EFFECTS

FOOD SCIENCE AND TECHNOLOGY

Additional books in this series can be found on Nova's website under the Series tab.

Additional e-books in this series can be found on Nova's website under the e-book tab.

FOOD SCIENCE AND TECHNOLOGY

LYCOPENE

FOOD SOURCES, POTENTIAL ROLE IN HUMAN HEALTH AND ANTIOXIDANT EFFECTS

JACOB R. BAILEY
EDITOR

nova publishers
New York

Library of Congress Cataloging-in-Publication Data

ISBN: 978-1-63117-927-3
Library of Congress Control Number: 2015931168

Published by Nova Science Publishers, Inc. † New York

CONTENTS

PREFACE

This book discusses food sources, the potential role in human health of lycopene, and the antioxidant effects lycopene possesses.

Chapter 1 – Carotenoids, naturally occurring pigments, are exclusively synthesized by plants and micro-organisms. Humans are not able to synthesize carotenoids and depend entirely on dietary intake of fruits, vegetables and nutritional supplements. The excessive generation of reactive oxygen species (ROS) induces "oxidative stress" that has been implicated in the disease causation and may be important progenitors in cancer progression. Lycopene, a red coloured carotenoid phytonutrient, naturally produced in highest concentration in tomato and tomato-based products is a potent $^1O_2$ quencher, and has been found to possess "antioxidant" and "antitumor" properties that mitigate the damaging effects of oxidative stress. In recent years, lycopene has gained more widespread attention as potential agent for prostate cancer prevention among other phytochemicals. Many studies have demonstrated the impact of lycopene in modulating the epigenome in diverse human cancers and in fetus. We summarize here the impact of antioxidant usage, using lycopene, to counteract the effects of oxidative stress-related diseases and for those at high risk of cancer and undergoing treatment and discuss whether the other dietary antioxidants are also having an effective therapeutic strategy. Furthermore, we also discuss the implementation of epigenetic-based therapies using lycopene alone or in combination with other plant-based diet on health care. Further elucidation will open the new avenue to identify the promising therapeutical potentials of novel antioxidants in treating and preventing disease.

Chapter 2 – A diet based on the consumption of fresh fruits such as tomato has been associated with health protection and longevity, due to their nutraceutical value. The latter is influenced by the presence of antioxidant compounds such as vitamin C (ascorbic acid), vitamin E (tocopherol), carotenoids, flavonoids as well as phenolic acids. Indeed, these molecules are able to neutralize reactive oxygen species (ROS) and, for this reason, are worldwide recognized as beneficial for preventing human diseases among which cancer and cardiovascular pathologies.

Lycopene is a natural pigment synthesized by photosynthetic plants and represents the most abundant carotenoid in tomatoes, being present in concentrations ranging from 9-150 mg/kg depending on the variety. The increase in dietary lycopene has been associated to a decreased risk of chronic diseases and one of the possible mechanisms involved in this action is represented by its antioxidant activity.

Lycopene as well as chlorophyll are lipophilic antioxidants, contributing to the antioxidant activity of tomato berries and leaves. However, the antioxidant activity of lipid extract from leaves was shown to be 45-70-fold higher than that from berries. This is because chlorophyll behaves as a fast lipophilic antioxidant (FLA), accounting for the 38% of FLA present in the lipid extract from leaves, whereas lycopene can be classified as a slow lipophilic antioxidant (SLA) responsible for the 40% of the antioxidant activity due to the slow fraction of the lipid extract from berries. Moreover, the presence of chlorophyll in a mixture containing lycopene increased the antioxidant activity compared to lycopene alone.

The role of chlorophyll for human health, for example in reducing risk of colon cancer, is more and more recognized by scientific community and its presence in food products such as not-fully ripe berries could substantially increase their nutritional value, in combination with high concentrations of lycopene.

Chapter 3 – Carotenoids are substances produced by fruits and vegetables, which are responsible for their red and orange tones. Lycopene belongs to the group of carotenoids and currently appears as one of the most potent antioxidants. Most carotenoids present linear structure with 40 carbons with 3 to 15 conjugated double bonds, which impart their property of absorbing light at wavelengths between 400 and 500 nm. Lycopene is a carotenoid without pro-vitamin A activity, lipid-soluble, composed of eleven conjugated bonds and two double bonds unconjugated. Lycopene is considered the carotenoid that has the highest scavenging capacity of singlet oxygen, possibly due to the presence of two double bonds unconjugated, which offers greater reactivity. For these reasons it is important to determine the amount of lycopene on different samples. Lycopene is the predominant carotenoid in human plasma and tissues, being found in a limited number of red colored foods as tomatoes and their products, guava and watermelon. The first step is the extraction of lycopene from the samples. This procedure can be made with different types of solvent and different kinds of extraction processes like liquid-liquid extraction (LLE), Solid-liquid extraction, dispersive solid phase extraction(d-SPE), solid phase extraction (SPE). The amount of lycopene after the extraction has been determined in food or biological samples by many analytical methods, such as UV -Visible spectrophotometry, liquid chromatography connected to electrospray-ionization (LC/ESI-MS), reverse phase liquid chromatography (RP-LC), supercritical fluid chromatography (SFC), matrix assisted desorption ionization (MALDI) and especially LC with spectrophotometric detection.

Chapter 4 – The lycopene ((6E,8E,10E,12E,14E,16E,18E,20E,22E, 24E, 26E)-2, 6, 10, 14, 19, 23, 27, 31-octamethyldotriaconta-2, 6, 8, 10, 12,14, 16, 18, 20, 22, 24, 26, 30-tridecaene) is part of the over 600 carotenoids group which exists in nature, the class of carotenes, due to its structure that is formed only of carbon and hydrogen. It is also one of the 25 carotenoids which can be found in human plasma and tissues. Its chemical formula is $C_{40}H_{56}$, and its atoms are arranged in a long chain acyclic, with 13 double bonds, and 11 of these bonds being conjugated bonds. Due to the absence of the β-ionone ring at the end of its molecule structure, it has no pro-vitamin A activity as other carotenes; it is lipo-soluble or insoluble in water, methanol and ethanol. It is a plant pigment with reddish color, being regarded as the carotenoid with the largest capacity of scavenging singlet oxygen and the ability to reduce the rates of free radicals, even being able to make them inactive. As a result of these aspects, it performs an important antioxidant action in the body and is associated therewith to research involving the prevention and treatment of different pathologies. Studies show that the activity of this carotenoid may reduce the occurrence of various types of cancer, such as esophageal, gastric, prostate and lung, and assist in the treatment of pancreatic cancer, colon and rectum, and it is inversely associated with the risk of myocardial infarction. Lycopene can be extracted from some fruits and vegetables, the main sources are tomato (*Lycopersicon esculentum*), red guava (*Psidium guajava*), watermelon (*Citrullus lanatus*), papaya (*Carica papaya*) and surinam cherry (*Eugenia uniflora*). The tomato (in its natural form) is the one with the highest concentration of lycopene in accordance with the degree of ripening, available as trans isomers (*trans*-lycopene) to be more stable in nature. However, the *cis* isomers of lycopene are most commonly found in plasma and absorbed by the human body, because they have greater solubility in micelles. This *cis-trans* isometry conversion is facilitated by thermal processing because it facilitates the rupture of the membrane, increasing the solubility and availability of lycopene. Therefore, for better absorption of lycopene in the body, it should be ingested cooked, like in tomato sauce and foods.

Chapter 5 – Lycopene is a natural red pigment mainly found in tomato, tomato products, and lycopene-carrots, which belongs to the group of carotenoids. It shows a very high quenching rate of reactive singlet oxygen, due to its apolar and acyclic molecular structure. Epidemiological evidence has suggested that lycopene can protect individuals from colorectal cancer and prostate cancer. However, as many members among the group of carotenoids, lycopene is very sensitive to heat, oxidation and light, due to their unsaturated chemical structures. It will limit the activity and potential health benefits of lycopene. Therefore, in order to improve the stability of lycopene, the processes for improving the stability of lycopene, must be developed and investigated, which will be mainly discussed in this chapter.

In: Lycopene
Editor: Jacob R. Bailey

ISBN: 978-1-63117-927-3

Chapter 1

POTENTIAL ROLE OF LYCOPENE AS ANTIOXIDANT AND IMPLICATIONS FOR HUMAN HEALTH AND DISEASE

Pushpinder Kaur[1] **and Jaspreet Kaur**[2]*
[1]Linus Pauling Science Center, Oregon State University, Corvallis, US
[2]University Institute of Engineering and Technology, Department of Biotechnology, Panjab University, Chandigarh, India

ABSTRACT

Carotenoids, naturally occurring pigments, are exclusively synthesized by plants and micro-organisms. Humans are not able to synthesize carotenoids and depend entirely on dietary intake of fruits, vegetables and nutritional supplements. The excessive generation of reactive oxygen species (ROS) induces "oxidative stress" that has been implicated in the disease causation and may be important progenitors in cancer progression. Lycopene, a red coloured carotenoid phytonutrient, naturally produced in highest concentration in tomato and tomato-based products is a potent $^1O_2$ quencher, and has been found to possess "antioxidant" and "antitumor" properties that mitigate the damaging effects of oxidative stress. In recent years, lycopene has gained more widespread attention as potential agent for prostate cancer prevention

* Corresponding Author: Pushpinder Kaur; Former Research Associate; Linus Pauling Science Center; Oregon State University; Corvallis, USA; E-mail: pushpinderkaur2006@gmail.com.

among other phytochemicals. Many studies have demonstrated the impact of lycopene in modulating the epigenome in diverse human cancers and in fetus. We summarize here the impact of antioxidant usage, using lycopene, to counteract the effects of oxidative stress-related diseases and for those at high risk of cancer and undergoing treatment and discuss whether the other dietary antioxidants are also having an effective therapeutic strategy. Furthermore, we also discuss the implementation of epigenetic-based therapies using lycopene alone or in combination with other plant-based diet on health care. Further elucidation will open the new avenue to identify the promising therapeutical potentials of novel antioxidants in treating and preventing disease.

Keywords: lycopene, oxidative stress, disease, cancer, epigenetics

ABBREVIATIONS

8-OHdG	8-hydroxy deoxoguanosine
AD	Alzheimer's Disease
ALP	Alkaline Phosphatase
ARMD	Age-Related Macular Degeneration
BMC	Bone Mineral Content
BMD	Bone Mineral Density
BP	Blood Pressure
CAD	Coronary Artery Disease
CAPE	Caffeic Acid Phenethyl Ester
CCAIMT	Common Carotid Artery Intima-Media Thickness
Cdk	Cyclin-dependent kinase
CHD	Coronary Heart Disease
COLIA1	Collagen Type I alpha 1
COX	Cyclooxygenase
CVD	Cardiovascular Disease
DBP	Diastolic Blood Pressure
DMRs	Differentially Methylated Regions
EGCG	Epigallocatechin Gallate
EGF	Epidermal Growth Factor
ER	Estrogen Receptor
FDA	Food and Drug Administration
GSH	Glutathione
HDL	High-Density Lipoprotein

HUVECs	Human Umbilical Vein Endothelial Cells
ICAM-1	Intercellular Cell Adhesion Molecule-1
IGF-1	Insulin-like Growth Factor 1
IGFBP-3	Insulin-like Growth Factor-Binding Protein 3
IVF	*In Vitro* Fertilisation
LDL	Low-Density Lipoprotein
lncRNAs	long non-coding RNAs
LOAD	Late-Onset Alzheimer's disease
MDA	Malonyldialdehyde
miRNAs	microRNAs
MMP-2	Matrix Metalloproteinase-2
MMPs	Matrix Mellaoproteinases
NF-kB	Nuclear Factor-kB
NMU	N-methyl-N-nitrosourea
OR	Odds-Ratio
OVX	Ovariectomized
PA	Protocatechuic aldehyde
PCNA	Proliferating Cell Nuclear Antigen
PCOS	Polycystic Ovarian Syndrome
PD	Parkinson Disease
PDGF	Platelet-derived Growth Factor
PDGFR-beta	Platelet-derived Growth Factor receptor beta
PGF2α	Prostaglandin F2α
PMCA3	Placental Calcium Transporter
ROS	Reactive Oxygen Species
SBP	Systolic Blood Pressure
SCO	Sertoli Cell Only
SNCA	α-synuclein
SOD	Superoxide Dismutase
T2DM	Type2 Diabetes Mellitus
TNF-α	Tumor Necrosis Factor-alpha
VCAM-1	Vascular Cell Adhesion Molecule-1
VDR	Vitamin D Receptor
VEGF	Vascular Endothelial Growth Factor

INTRODUCTION

Environmental risk factors, lifestyle (including dietary factors) have a significant effect on chronic diseases such as obesity, diabetes, cardiovascular, dental disease, osteoporosis, and cancer are a major burden to the society and is the leading cause of death worldwide [1]. Genetic and epigenetic factors are also the major contributors to the human disease. "Reactive oxygen species" (ROS) are generated by internal factors through normal metabolic activity and from external factors (pollutants, tobacco, xenobiotics, exercise, diet and radiation) [2, 3]. ROS such as superoxide, peroxyl radical, hydrogen peroxide, hydroxyl radical and peroxynitrile react with cellular components, resulting in oxidative stress, eventually leading to several human diseases and aging [4-8].

Antioxidants, also known as "free radical scavengers", are man-made and natural substances that limit oxidative stress by donating an electron to the free radical and neutralizing it, thus protecting the cells from damage [9]. The antioxidants produced within the body are glutathione peroxidase, glutathione reductase, superoxide dismutase, catalase, thioredoxin reductase, heme oxygenase, biliverdin reductase, ubiquinol and uric acid [10] and other lighter antioxidants such as vitamin E, vitamin C, carotenoids and polyphenols [11] are available in diet. Carotenoids, the natural pigments synthesized by plants, algae and photosynthetic bacteria, are powerful antioxidants associated with reduced risk of several chronic health disorders including cancer [12-14]. The main six carotenoids found in human serum are α-carotene, β-carotene, β-cryptoxanthin, lycopene, lutein and zeaxanthin. Recently, lycopene has gained much attention for its preventive properties against chronic diseases and cancer.

Lycopene, a fat-soluble pigment, has been found to possess antioxidant and antitumor properties. Epidemiological studies suggest that higher intake of lycopene reduces the risk of cancer [15] and cardiovascular diseases [14]. Lycopene appears to exert its protective effect through several biochemical mechanisms involving antioxidant activity such as quenching of singlet oxygen and reduction of free radicals, antimetastatic activity, apoptosis, cell cycle arrest, cell-cell communication, gene function regulation, carcinogen metabolism and metabolic pathways involving phase I and phase II drug-metabolizing enzymes. Recently, studies provide key evidence that this plant-based dietary agent impact epigenetic marks, altering normal epigenetic states as well as reversal of abnormal gene activation or silencing [15, 16], and can act as a putative nutraceutical to be incorporated into an "epigenetic diet". In this review, we discuss the protective effect of lycopene on oxidative stress-related chronic diseases, including cancer. Dietary antioxidants, with special reference to lycopene will be discussed, that targets the epigenome in reducing the risk and provide new insight into clinical research for better disease prevention and therapy.

DIETARY SOURCES, CHEMISTRY AND ABSORPTION OF LYCOPENE

The highest source of lycopene in the typical diet is tomato, providing at least 85% of all the dietary lycopene. Although other food sources such as guava, cranberry, peaches, pink grapefruit, apricot, watermelon and papaya also provide lycopene but in smaller amount [17]. The most readily absorbed sources of lycopene are the processed tomato products such as pasteurized tomato juice, soup, salsa, ketchup, spaghetti sauce, etc. and many laboratories have quantified its content.

Lycopene is an acyclic non-polar carotenoid having the molecular formula $C_{40}H_{56}$, contains 11 conjugated and 2 unconjugated double bonds. Its structure lacks terminal β-ionone ring and pro-vitamin A activity. As a polyene, it can undergo isomerization to cis-trans isomers. These isomers are induced by light absorption, heat exposure and chemical reaction. Lycopene exists predominantly as 80-97% in all-trans (E), thermodynamically stable configuration in tomatoes, fruits and vegetables [18]. In human plasma and body tissues, lycopene is found as a mixture of 50% cis (Z) and 50% all-trans isomers [19].

Lycopene is more easily absorbed from cooked tomato products, rather than raw, because of breakdown of the cell matrix by heat which makes the absorption of lycopene more efficient. Heat induces isomerization of lycopene from all trans to cis configuration and makes it more bioavailable [20]. The cis isomers of lycopene prevail in the human serum, improve the absorption in the gastrointestinal tract and improve distribution of lycopene throughout the body may occur by the blood and other fluids. Due to its fat-soluble nature, the absorption is further enhanced when lycopene-rich vegetables and fruits are consumed with oil [21, 22]. The digestion and absorption of dietary lycopene is a complicated process that consists of incorporation of lycopene from food matrix into micelles, packaging into chylomicrons, transportation to the tissues such as the liver, adrenal glands, testes, prostate and lungs [23]. The ingested lycopene is then oxidized and metabolized in the body. Recent studies have shown that apo-10' lycopenoic acid, metabolite of lycopene is an effective chemopreventive agent against hepatic tumorigenesis, inflammation [24] and lung cancer [25]. Lycopene supplement pills also play a key role in the disease prevention.

Mechanism of Action

Although numerous studies have attempted to explore the mechanistic process, however, the biological mechanism(s) of action of lycopene are still under investigation. Oxidative stress induced by ROS causes damage to DNA and macromolecules, has been implicated in the causation of many diseases [26-31]. Lycopene has robust anti-oxidant defense system, attributed to its acyclic structure, numerous conjugated double bonds, high hydrophobicity and thus prevents the onset of carcinogenesis and atherogenesis processes by protecting/stabilizing biomolecules such as DNA, proteins, lipids and lipoproteins [32]. Lycopene possesses the ability to quench singlet oxygen and, like all other dietary carotenoids, also scavenges the free radicals *via* three different mechanisms: adduct formation, electron transfer and hydrogen atom transfer [33-35]. Galano et al. [36] reported that lycopene and torulene are highly reactive scavengers, than β-carotene, of peroxide radicals. Bast et al. [37] also revealed that lycopene acts as an antioxidant by repairing radical traps generated from vitamin E and C and effectively reduces α-tocopheryl radical.

Growth Factors

Lycopene supplementation also modulates several growth factors such as insulin-like growth factor 1 (IGF-1), vascular endothelial growth factor (VEGF), epidermal growth factor (EGF) and platelet-derived growth factor (PDGF) that play a pivotal role in tumor development and metastasis. Elevated serum levels of IGF-1 have been implicated as a possible risk factor of various human cancers [38]. However, the studies conducted on the influence of lycopene supplementation on the serum IGF-1 have been somewhat controversial. Walfisch et al. [39] conducted the trial study on lycopene supplement (Lyc-O-Mato capsules; 30mg/day) and observed 25% decrease in serum IGF-1 in colon cancer patients. The positive effect of lycopene on IGF-binding protein 3 (IGFBP-3) concentrations were observed in ten healthy men after supplement intake of 15mg/day for 4 weeks [40]. Another study by Riso et al. [41] revealed the reduction in serum IGF-1 in 20 healthy persons who consumed Lyc-O-Mato drink (5.7mg/day) but without any change in IGFBP-3 level. Likewise, Vreiling et al. [42] conducted the large dietary intervention study on 71 persons having family history of colorectal cancer and observed no effect of Lyc-O-Mato capsule (30mg/day) on serum IGF-1 and IGFBP-3. Lycopene treatment also delayed G1-S transition and reduced growth stimulation of mammary cancer cells by IGF-1 [43]. Wu et al. [44] observed that lycopene inhibits PDGF-BB signaling and cell migration in human cultured skin fibroblasts. Chiang et al. [45] investigated the attenuation of phosphorylation of PDGF receptor beta (PDGFR-beta), extracellular signal-regulated kinase 1/2 (ERK1/2), p38, and c-Jun N-terminal kinase (JNK), induced by PDGF-BB, with the increasing concentration of lycopene. Anti-VEGF monoclonal antibodies such as ranibizumab, bevacizumab and pegaptanib, approved by Food and Drug Administration (FDA), are being used to treat colorectal, non-small-cell lung cancer and age-related macular degeneration (ARMD). Although, anti-VEGF agents are not effective in all patients [46], however, dietary antioxidants such as flavanoids, lutein, zeaxanthin and lycopene afford a remarkable protection against ARMD [47]. Lycopene or β-carotene supplementation strikingly reduced the plasma levels of matrix metalloproteinase-2 (MMP-2) and VEGF and inhibited the tumor growth in the lung of the nude mice injected with human hepatoma cells [48]. Oral supplementation of low and high dose of lycopene (4 and 16mg/kg) and β-carotene (16mg/kg) reduced tumor growth and decreased the expression of proliferating cell nuclear antigen (PCNA) in prostate tumor. Furthermore, high dose of lycopene also reduced elevated plasma levels of VEGF [49]. Chen et al. [50] determined the anti-angiogenic action of lycopene both *in vitro* and *in vivo* by inhibiting MMP-2/uPA system through VEGFR-2 mediated PI3K-Akt and ERK/p38 signalling pathways. These studies help in understanding the "nutraceutical" properties of lycopene. Despite thorough investigations, the results have been somewhat contradictory on the impact of lycopene supplements on IGF levels. The fundamental key question arises is whether these supplement pills are safe? If yes, whether these supplements have an ability to "prevent" or "reverse" the lesions? Whether these supplements act on multiple cellular pathways? Additional research is required to answers these questions, in order to determine the molecular basis, potential efficacy, safety and optimal dosing that may impact cancer prevention and treatment.

SIGNALING PATHWAYS

There is mounting evidence that lycopene inhibits cancer cell growth and cell cycle, which in turn, can influence different signaling pathways such as androgen signaling, cytokine (IL-6, IL-4) and growth factor signaling (e.g., IGF, Wnt/beta-catenin), PI3K/Akt, MAPK, and Jak/STAT pathways. Tang et al. [51] reported that lycopene inhibits human colon cancer cell proliferation with a IC_{50} value of 10 μM *via* controlling the Akt signaling pathway and non-phosphorylated β-catenin protein level. The authors have also shown that lycopene reduces promoter activity and regulate the protein expression of cyclin D1 in human colon cancer cells. Dietary administration of lycopene in combination with vitamin E and selenium had a significantly lower risk of prostate tumor in the *Lady* transgenic mouse model than vitamin E and selenium alone, which revealed the promise of plant-based dietary agents for cancer prevention [52].

The same group investigated the effect of Capsaicin, a major biologically active ingredient of chili peppers, alone or in combination with lycopene and revealed that "combinatorial therapy" exhibits anti-proliferative and pro-apoptotic effects and also induces cell cycle arrest in human prostate cancer cell lines. The treatment also altered the protein expression of cleaved PARP, Caspase-3 and Bax/Bcl-xL, involved in apoptosis signaling pathway [53]. Lycopene was shown to affect Ras and NF-kB activation by inhibiting ROS production [54]. Lycopene interferes with MAPK signaling pathway in breast cancer cells [55] and inhibits HMG-CoA reductase expression in prostate PC-3, colon HCT-116, HT-29 and lung BEN cancer cells [54].

Lycopene was reported to inhibit the genotoxicity and mutagenicity of H_2O_2 and showed a reduction of DNA damage in human hepatoma HepG2 cells [56]. These studies suggest that lycopene exhibits its anticancer activity through interaction with multiple cell signaling pathways and also helps in understanding that lycopene alone, or in combination with other dietary agents is effective in combating the cancer. These findings could have a significant impact on understanding of molecular mechanism of action of lycopene against cancer and will lead to identify the candidate biomarkers for therapeutic innovation.

Cell Cycle, Cellular Invasion and Apoptosis

The dietary phytochemical lycopene is active at different stages of cancer progression by interfering with signaling pathways, cell cycle arrest, apoptosis and inhibiting angiogenesis and metastasis. Cell cycle is tightly regulated by cyclin and cyclin-dependent kinases (Cdk). Any alteration in the cell cycle checkpoints results in increased cellular proliferation which further leads to the progression of disease [57]. Apoptosis is physiological process controlled by a series of cellular events inducing programmed cell death. Dysregulation of apoptosis has been linked to several factors such as nuclear factor-kB (NF-kB) and AP-1. A number of epidemiological and clinical trial studies suggest an antioxidant potential and protective effect of lycopene against disease, however, the clear molecular mechanism, including the induction of apoptosis and cell cycle arrest is still not fully understood [58, 59]. Interestingly, lycopene has been shown to have growth inhibitory effects at 1μM concentration on human prostate cancer cells and at 5μM, the percentage of cells increased in the G2/M phase from 13 to 28% and decreased from 45 to 29% in S-phase, accompanied by an increase of apoptotic cells [60]. Lycopene reduced the expression of cyclin D1, E, Cdk2 and 4 and also inhibited the activation of IGF-1 receptor in breast cancer cells with subsequent delay in cell cycle progression [43]. In contrast, Burgess et al. [61] observed no reduction in cell proliferation in five lycopene treated human cancerous and non-cancerous cell lines (A431, skin carcinoma; DU-145, prostate carcinoma; HS-68, noncancerous skin; A549, lung carcinoma; and HS-578T, breast carcinoma cells) from 0.0001-10μM, however, cell proliferation was reduced only in HepG2 at 1 and 10μM. At high dose, the HepG2 (liver adenocarcinoma cell line) and IMR-90 (non-cancerous lung cell line) showed a decreasing trend in cell number. These studies suggest that the anticancer effects of lycopene are cell-type or tissue-type specific and much research should be focused to explore the underlying mechanism of action of lycopene.

Therapeutic Effects of Lycopene on Disease and Cancer

The induction of oxidative stress by ROS leads to cell transformation and has been implicated in several acute and chronic diseases. To combat these diseases, nutritional guidelines suggest increased intake of dietary antioxidants such as vitamins C and E, selenium, beta-carotene, lycopene, lutein, and zeaxanthin in the diet. No recommendations have been made for the routine consumption of antioxidant supplements for the prevention and treatment of chronic diseases [62]. A number of chromatin modifying drugs and naturally occurring phytochemicals that have the ability to modulate the "epigenome" in view of therapeutic purposes for several human diseases are under consideration. Epigenetic intervention using dietary compounds is now an emerging area of research and due to its "reversible" character, DNA methylation machinery, chromatin modifications and microRNAs (miRNAs) represent a promising target for therapy. A number of studies have revealed that natural dietary agents targeting epigenetic pathways have protective effects in preventing diseases, including cancer. However, there are many as-yet undiscovered agents that might play an important role in multi-aspect cross-talk among distinct epigenetic pathways. The present chapter will provide an insight on the role of dietary lycopene in various human disease and cancer and also discuss its effect on epigenome.

CARDIOVASCULAR DISEASE (CVD)

Oxidative modification of low-density lipoprotein (LDL) has been reported to play a major role in artherosclerosis and coronary heart disease and LDL-associated antioxidants such as lycopene and vitamin E provided effective barrier against oxidation [63-65]. Conversely, β-carotene gave ambiguous results in inhibiting LDL oxidation. There is now accumulating evidence from epidemiological and human intervention studies that "lycopene-rich diet" provides protection against coronary heart disease (CHD).

In a multicenter case-control study (EURAMIC), participants were recruited from 10 European countries to assess the association between adipose tissue antioxidant status and acute myocardial infarction. After infarction, adipose tissue samples were collected to determine α-, β-carotene and lycopene levels. Interestingly, men with highest intake of lycopene demonstrated a 48% reduction in the risk of myocardial infarction [66]. Rissanen et al. [67] evaluated the relationship between low plasma lycopene concentrations and common carotid artery intima-media thickness (CCAIMT) in subjects with early artherosclerosis of eastern Finland. Recently, Zou et al. [68] have suggested that the lycopene and lutein combination therapy was more effective for reducing CCAIMT in Chinese subjects.

Many *in vivo* studies have identified the possible link between epigenetic modifications and risk of CVD [69-71]. Sharma et al. [72] revealed significant association between plasma homocysteine levels (known independent risk factor for heart disease) with global DNA methylation in coronary artery disease (CAD) patients. The same group also identified the 72 hypermethylated differentially methylated regions (DMRs) in patients with CAD [73]. Hypermethylation of the promoter region has been demonstrated for several genes in artherosclerosis such as superoxide dismutase, estrogen receptor alpha, estrogen receptor B-gene, *FOXP3* and endothelial nitric oxide synthase [74-78]. Cash et al. [79] identified the *LINE-1* methylation in CVD associated with altered levels of LDL and high-density lipoproteins (HDL) in Samoan Islanders. Many studies have established a strong link between chromatin modifications and artherosclerotic plaques [80-85]. The signatures of the non-coding component of the genome, including miRNAs and long non-coding RNAs (lncRNAs), have also been functionally identified as regulators in artherosclerosis [86-89]. Various studies have provided evidence suggesting the use of dietary antioxidants such as sulforaphane and curcumin, having an ability to modulate the epigenome, reduced the risk of artherosclerosis and CVD [90-97]. Wu et al. [98] observed that diet containing sulforaphane reduced the oxidation and risk of CVD in hypertensive stroke-prone rats. Sulforaphane has also been shown to reduce inflammation by suppressing the inflammatory markers i.e. vascular cell adhesion molecule-1 (VCAM-1) in endothelial cells through activation of Nrf2 [90, 91, 94]. Protocatechuic aldehyde (PA), act as HDAC inhibitor, downregulates the expression of VCAM-1 and intercellular cell adhesion molecule-1 (ICAM-1) in human umbilical vein endothelial cells (HUVECs) and prevents the development of artherogenesis [99]. Accumulating evidences have shown that dietary agents have the potential to influence the epigenetic state and also in preventing and reducing the risk of heart-related diseases. The recognition of lycopene in CVD epigenetics has not been yet clearly established. As the epigenetic diet has intriguing potential, other dietary agents that are still untested should also be considered to check their therapeutic potential against CVD.

Diabetes

Increase in the oxidative stress leads to insulin resistance, β-symbol dysfunction, glucose and fatty acid overload which further results in the onset of diabetes. In recent years, research has also focused on the contribution of epigenetic changes that occur during the development of diabetes. Hyperglycemia results in altered recruitment of transcriptional machinery by inducing epigenetic alterations in vascular cells [100, 101]. Ling and colleagues [102] observed the DNA methylation in *PPARGC1A* promoter, in pancreatic islets of patients with type 2 diabetes. Indeed, changes in the histone modification pattern were also observed in the insulin gene in human islet-derived precursor cells [103]. Diabetes induces *in vivo* chromatin remodeling, stimulates histone modifications and the recruitment of NF-kB to the promoters of inflammatory genes also [104]. Hyperglycemia enhances the expression of inflammatory genes, induces aberrant histone K4 methylation and dysfunction of LSD1 [105, 106]. Various studies have identified a series of dysregulated miRNAs that are implicated in the pathogenesis of diabetes and could be used as potential predictors for diagnosis and treatment [107].

Consumption of dietary antioxidants could potentially prevent and delay the complication of diabetes [108-110]. Upritchard et al. [111] revealed that dietary supplement of tomato juice increased plasma lycopene levels to be associated with the reduced risk of myocardial infarction in diabetes. Lycopene also acted as an ACE inhibitor in *in vitro* study [112]. Neyestani et al. [113] investigated the immune-enhancing activities of lycopene by inhibiting MDA-LDL formation and prevention of ox-LDL uptake by macrophage in a double-blind placebo-controlled clinical trial study of 35 patients with type2 diabetes mellitus (T2-DM). On the contrary, Wang et al. [114] conducted the cohort study of more than 35,000 women having T2-DM and investigated that intake of lycopene or lycopene-containing food was having no significant association with diabetes. No clear relationships were found with the other antioxidants supplementation, including vitamin E and C, with any improvement of glycaemic control in patients with T2 diabetes [115]. However, Montonen et al. [116] revealed that intake of total vitamin E, α-tocopherol, β-tocopherol, γ-tocopherol and β-cryptoxanthin reduced the risk of development of T2 diabetes.

Many studies have shown that altered epigenetic marks exist in the diabetes and research efforts have also been focused on the prevention of diabetes by various dietary antioxidant supplementation, including lycopene, however, the research in this field is still curtailed as the questions regarding the mechanism of conferring this protection and synergism of lycopene with other dietary factors are still unanswered.

BONE HEALTH

Persistent production of ROS causes oxidative stress, inflammation and other chronic diseases including osteoporosis, bone tumor development, diabetes-induced bone complications and joint inflammatory diseases. Many researchers observed that oxidative stress can influence bone remodeling [117-119]. Various dietary antioxidant therapies, including lycopene, have shown promise for improving the effectiveness of bone health. Positive correlation was found between lycopene intake and bone strength in young female rats [120]. Liang et al. [121] reported that increased consumption of lycopene for eight weeks results in decreased serum Ca, P, alkaline phosphatase (ALP), and IL-6 concentration, and enhances serum estrogen level, bone mineral density (BMD), and bone mineral content (BMC) in ovariectomized (OVX) rats. Despite these observations, the actual molecular mechanism of action of lycopene in the role of oxidative stress in bone health is not well understood. There is growing evidence that other dietary agents such as flavanoid, hesperidin [122] and pomegranate [123] also play a fundamental role in preventing the development of bone loss induced by ovariectomy in mice. Orange juice or grapefruit juice also appeared to be particularly effective in bone protection against osteoporosis in senescent male rats [124].

Epigenetic processes have also been implicated in impaired bone health. *In utero* exposure to low maternal calcium intake results in lower bone mass in fetuses and neonates. Epigenetic alteration in response to *in utero* nutritional and environmental factors also deregulates histone modifications and alters the CpG methylation thus playing an important role in fetal bone development. Dietary intake of Vitamin D and calcium during pregnancy and lactation influences fetal skeleton development. Martin et al. [125] revealed that placental calcium transporter (*PMCA3*) gene expression influence neonatal whole-BMC. Other genes i.e. vitamin D receptor (*VDR*), the collagen type I alpha 1 (*COLIA1*) and estrogen receptor gene (*ER*) alpha also play pivotal role in bone formation, resorption as well as in regulating BMD [126]. The lycopene content in the maternal diet has also been proven to prevent inadequate growth of the fetus [127]. This helps in understanding how the maternal food consumption during the critical and sensitive period of development can have long-term effect on fetal health.

Dietary supplementation of the maternal diet with tomato juice had also counteracted the negative effects of nicotine on the lungs of the offspring [128]. Although the research on lycopene as a preventive agent is very promising, however, it remains to be seen whether lycopene confers the protection to the fetus during gestational and lactation exposure. Other challenges regarding the "ideal dosage", the "combinatorial therapy" of lycopene with other dietary components to improve the fetal bone development and whether the transmission of lycopene across multiple generations would be able to prevent successive generation from diseases and cancer, still needs to be addressed.

MALE INFERTILITY

Exposure to environmental factors such as chemicals, pesticides, heavy metals, electromagnetic radiations [129-131]; use of exogenous substances i.e. cigarette smoke, alcohol, testosterone supplements; chronic diseases i.e. obesity, stress, urogenital trauma and inflammation in the male reproductive system [132-134] contributes to male infertility. ROS are also involved in numerous physiological functions of sperm including capacitation, acrosome reaction and sperm motility [135]. Excessive ROS generation damages the sperm membrane, which induces apoptosis, low sperm count, motility impairment and damage to lipid, protein and DNA [136, 137].

Epigenomic abnormalities have also been linked to male infertility. Marques et al. [138] observed elevated methylation levels in eight moderate (17%) and fifteen severe (30%) oligozoospermic patients in H19 gene, suggesting a link between abnormal genomic imprinting and hypospermatogenesis. Benchaib et al. [139] investigated the impact of DNA methylation, using 5-methyl-cytosine immunostaining in semen samples from men undergoing *in vitro* fertilisation (IVF), on fertilization and implantation. The authors observed that high methylation level in human spermatozoa is associated with fertilization and embryogenesis. Another study by Marques et al. [140] explored distinct epigenetic variations in several disease-related genes (*BRCA1, BRCA2, HD, DM1, PSEN1* and *PSEN2*) and age-related genomic methylation in another subset of genes (*EED, CTNNA2, CALM1, CDH13*, and *STMN2*) within the germline of normal males. In addition, global increase in histone H4 acetylation machinery was also associated with Sertoli cell only (SCO) syndrome in infertile men [141].

Recently, antioxidant supplements have gained a lot of attention to promote health and to prevent and cure infertility. Numerous studies have been conducted to evaluate the effect of antioxidants (vitamin C, vitamin E, selenium, zinc, glutathione, L–carnitine and N–acetyl–cysteine) in the improvement of sperm quality and pregnancy rates in infertile subjects [137,

142-145]. Lycopene supplementation from various sources (tomato, synthetic or placebo) is found to be effective in patients against oxidative stress and in reducing lipid peroxidation [146]. Devaraj et al. [147] documented that a dose of 30mg lycopene/day significantly decreases DNA strand breaks and urinary 8-hydroxy deoxoguanosine (8-OHdG), biomarker of oxidative stress. Lycopene-enriched extenders improved the quality of turkey semen and protects the viability, osmotic resistance and DNA integrity during cryopreservation [148]. Lycopene supplementation (2000 mcg, twice a day for three months) showed a substantial benefit of 66% improvement in sperm concentration, 53% in sperm motility and 46% in sperm morphology in patients with idipathic infertility [149]. Other studies have also reported a significant increase in sperm count with lycopene supplementation [150, 151].

Lycopene is found in high concentration in testes, adrenal glands and prostate. Although, the available literature suggests that the average daily intake of this phytonutrient improves the fertility and spermatogenesis, however, additional research is needed on human sperm in order to determine the long-term impact of dietary antioxidants on successive generations.

Female Infertility

Oxidative stress response has been extensively studied on women's fertility and its substantial increase leads to acute complications such as pre-eclampsia, hyatidiform mole, polycystic ovarian syndrome (PCOS), endometriosis, abortions and unexplained infertility. Kuscu et al. [152] observed increased malonyldialdehyde (MDA) and superoxide dismutase (SOD) levels in PCOS subjects as compared to control. Due to increased production of ROS, prostaglandin F2α (PGF2α) level in endometrium is stimulated through cyclooxygenase (COX). SOD and PGF2α expression induces the variation in endometrium resulting in endometrial breakdown and shedding at menstruation [153, 154]. Induction of oxidative stress leads to aberrant angiogenesis and abnormal uterine bleeding [155]. Endometriosis is a gynaecological disorder where the endometrium grows outside the uterine cavity that involves increased expression of matrix metalloproteinases (MMPs). Interestingly, the expression of *MMP-9* was markedly reduced by curcumin treatment in endometriotic tissues. Furthermore, curcumin pretreatment reduces the expression of tumor necrosis factor-alpha (*TNF-α*) and also prevent lipid and protein peroxidation in tissue samples [156]. Two retrospective studies [157, 158] tested the potential effectiveness of melatonin in reducing *MMP-9* expression and peroxidation systems against endometriosis in *in vitro* mouse model. Melatonin markedly restored the reduced MDA level and increased SOD and catalase activity [159], supporting the notion that melatonin may also play a useful role in preventing endometriosis. Nahum et al. [160] revealed that lycopene inhibits cell cycle progression in human breast and endometrial cancer cells *via* reducing the cyclin-dependent kinase activities. Levy et al. [161] investigated that lycopene at half-maximal inhibitory concentration of 1-2 μM inhibits the proliferation of human endometrial, mammary and lung cancer cells in comparison to α- and β-carotene. Another study on rat model revealed that caffeic acid phenethyl ester (CAPE) treatment resulted in significant reduction in MDA concentrations and the levels of antioxidant enzyme activity [159]. Xu et al. [162] demonstrated that, when compared, both epigallocatechin gallate (EGCG) and vitamin E treatment on endometrial lesions, EGCG supplementation mediated downregulation of *VEGF-A* and reduced the lesion size as compared to vitamin E. Fambrini et al. [163] identified the HOXA10 methylation markers in the pathogenesis of ovarian endometriomas that could be used for prognosis and potentially therapy. Interestingly, folic acid treatment revealed an ability to decrease the promoter methylation of HOXA10 gene in pregnant women with endometriosis [164]. Direct readouts of genome-wide methylation array revealed hypermethylation in 59 genes and hypomethylation in 61 genes in the endometrium [165]. Numerous studies indicated that the aberrant histone modifications i.e. hyperacetylation of histones located in the promoter lesions of G-protein-coupled estrogen receptor 1, steroidogenic factor-1, and hypoxia-inducible factor-1 alpha genes; and hypoacetylation of histones located in the promoter lesions of estrogen receptor alpha, homeobox A10, CCAAT/enhancer-binding protein alpha, *p16(INK4a)*, *p21(Waf1/Cip1)*, *p27(Kip1)*, checkpoint kinase 2, death receptor 6, and E-cadherin genes play a major causal role in endometriosis [166]. These findings support the concept that epigenetic modifications can influence the risk of endometriosis.

Preeclampsia is dangerous complication of human pregnancy and the contributing factor to maternal and fetal/neonatal morbidity and mortality. Chappell et al. [167] demonstrated that the supplementation of antioxidants vitamin E and C at 16-22 weeks of pregnancy resulted in a reduced incidence of preeclampsia. In contrast, another study failed to demonstrate any effect of vitamin E and C supplementation in preventing preeclampsia [168]. Numerous clinical and randomized controlled trials of dietary supplements (fish oil supplementation, magnesium, calcium) revealed no beneficial results for preeclamsia [169]. A randomized controlled trial conducted among 251 primigravida women received either placebo or 2mg of lycopene twice daily may manifest *effective* protective action against preeclampsia and prevent inadequate growth of the fetus [127], but the underlying mechanism is still unclear. So far, very few studies have looked at the effect of lycopene on female infertility. Further randomized human controlled nutrition intervention trials of lycopene with well-designed parameters, including subject selection, synergistic interaction of lycopene with other phytonutrients, long-term follow up with emphasis on the mechanism of action are needed.

NEURODEGENERATIVE DISEASES

Oxidative stress plays a critical role in neurodegenerative diseases, such as Alzheimer's, Parkinson and Huntington's. Epigenetic mechanisms are believed to have an impact in the onset and progression of neurodegenerative disorders. Chouliaras et al. [170] observed the decrease in the hippocampal levels of the 5-methylcytidine (5-mC; 19.6%) and 5-hydroxymethylcytidine (5-hmC; 20.2%) markers in patients with Alzheimer's disease (AD). Bakulski et al. [171] sheds light on the molecular mechanism involved in late-onset Alzheimer's disease (LOAD) using genome-wide DNA methylation array in which discordant DNA methylation were observed in LOAD cases. Besides DNA methylation, the patterns of histone modification including low level of H3 acetylation [172] and increased level of HDAC6 and HDAC2 were seen in post-mortem AD brain [173, 174]. DNA methylation signatures were observed in a number of studies in Parkinson disease (PD) genes in post-mortem brain and peripheral blood cells [175]. Methylation analysis of several brain regions of sporadic PD patients revealed reduced methylation levels of the gene coding for α-synuclein (SNCA) [176, 177].

Reports indicate that consumption of dietary antioxidants provide substantial protection against neurological disorders and may be incorporated into our daily diet. In a randomized controlled trial study, dietary intervention with selegiline hydrochloride or vitamin E leads to the reduction of AD [178]. Epigenetic intervention using SAM supplementation reduces the oxidative stress [179], reverses the altered methylation state of *PSEN1* promoter, leading to reduced progression of AD caused by vitamin D deficiency in mice [180-182]. Masaki et al. [183] associate the intake of vitamin E and C consumption with a lower prevalence of vascular and mixed dementias in 3385 men. Conversely, no significant beneficial effects were observed on AD. Studies on lycopene intake have been proved beneficial in lowering the risk of insulin-signaling deficits, oxidative stress, neuroinflammation and cognitive impairment in fructose-drinking insulin resistant rats [184]. Alzheimer's disease patients associated with vascular comorbidities had evidence of circulating antioxidant micronutrients i.e. retinol, vitamin E, lutein, zeaxanthin, lycopene and β-carotene depletion in their plasma levels [185]. In 3-nitropropionic acid-induced Huntington's disease rats, lycopene treatment at 2.5, 5 and 10mg/kg have shown to enhance the activities of biochemical and mitochondrial enzymes. In addition, l-arginine (50mg/kg) and L-NAME (10mg/kg) pretreatment with lycopene (5mg/kg) appears to have significant protective effect in 3-nitropropionic acid-treated rats [186]. Another study by the same group on 3-nitropropionic acid-treated morris water maze model, lycopene along with EGCG significantly improved memory and enhanced the activities of glutathione system [187]. The beneficial effect of lycopene supplementation in rotenone induced PD model increased MDA levels and enhances the activities of SOD and glutathione (GSH), suggesting its relevance for neuroprotection [188, 189]. Besides this extensive research, the epigenetic effect of lycopene in relation to brain diseases has not been yet elucidated. Understanding the nutritional epigenomics and its influence on humans will help to develop novel complementary therapeutic intervention for neurodegenerative diseases.

CANCER

Evidence suggests that intake of lycopene might influence the risk of various cancers i.e. prostate, breast, lung, leukemia and colon [190, 191]. In the Health Professionals Follow-up Study, Giovannucci et al. [192] documented that lycopene intake reduces prostate-related cancer risk (Relative Risk-0.79), although the tomato-sauce had strongest protective anti-cancer effect. More interestingly, the combinatorial intake of tomatoes, tomato sauce, tomato juice, and pizza is accompanied by a greater reduction in the risk of prostate cancer (Relative Risk-0.65). In another prospective US cohort study, amongst the various pigments, intake of carotenoids, the α-carotene and lycopene have a significant impact in reducing lung cancer [193]. Levi et al. [194] investigated the relationship between dietary intake of selected micronutrients and breast cancer risk. The adjusted odds-ratio (OR) tends to decrease for dietary vitamin C (0.42) and lycopene (0.43) intake and showed an inverse association with the risk of breast cancer. Furthermore, in their statistical model including a continuous term for the seven micronutrients, lycopene and vitamin C showed a continuous trend towards greater decrease of breast cancer risk. Dietary intake of tomato and tomato-based products were associated with 31% reduced risk in pancreatic cancer [195].

An increasing body of evidence has shown the deregulation of epigenetic mechanism in many cancer types. The dietary agents can modulate the altered epigenome by restoring the expression and function of epigenetically silenced genes. Lycopene at 10 μM regulates the expression of genes involved in cell cycle regulation, DNA repair and apoptosis in MCF-7 and MDA-MB-231 breast cancer cells and in MCF10A cells [55, 196]. Until now, only three studies have investigated the ability of lycopene to target epigenetic mechanism in breast and prostate cancer. King-Batoon et al. [16] have shown that lycopene at 2 μM alters *GSTP1* in MDA-MB-468 cells, *RARβ* and *HIN-1* gene expression in non-cancer fibrocystic breast cells by demethylating the promoter region. In contrast, no effect was observed in response to lycopene and its putative metabolite apo-10′-lycopenal on *GSTP1* methylation in human prostate cancer LNCaP cell line [197]. However, lycopene treatment alters the *GSTP1* methylation in correspondence with mRNA and protein expression in an androgen-independent prostate cancer PC-3 cell line [198]. Further detailed research is required to understand the mechanism of lycopene on epigenetic modulation on gene expression. Both large-scale and long-term human intervention studies should be designed specifically to determine the gene-diet interactions in cancer prevention.

WHOLE FOOD *VS.* LYCOPENE SUPPLEMENTATION

Several dietary intervention trials have compared the protective effect of lycopene supplements and lycopene-rich food against human disease; however the results are somewhat contradictory. Recently, Gajendragadkar et al. [199] have reported that commercially available oral lycopene (Ateronon, Cambridge Theranostics, UK) increased the blood vessel function in patients suffering from heart disease. Other supplements such as Lyc-O-Mato® and LycoRed® (both manufactured by LycoRed Natural Products Industries Ltd., Beer-Sheva, Israel) contains 6-7% lycopene and are available in soft gel capsule. Boileau et al. [200] studied the effect of freeze-dried whole tomatoes (tomato powder; 13mg lycopene/kg) and a diet containing pure synthetic lycopene (161mg lycopene/kg) in a N-methyl-*N*-nitrosourea (NMU)-androgen rat carcinogenesis model. Mortality from prostate cancer was reduced by 26%

in tomato powder group and in pure lycopene group only a 9% risk was reduced as compared to control. A number of studies observed decrease in blood pressure (BP) with daily oral supplementation of tomato extract or tomato juice [201-205]. Ried et al. [206] found that lycopene treatment significantly decrease only the systolic blood pressure (SBP), not the diastolic BP (DBP). Paterson et al. [207] reported increase in BP with lycopene (4.5mg/day, 4 weeks) while others found no effect on BP [111, 208, 209]. Other carotenoids in tomatoes, in addition to lycopene, are phytoene, phytofluene, zeta-carotene, gamma-carotene, beta-carotene, neurosporene, and lutein. Although, several dietary research findings have claimed that β-carotene diet-based regime have protective effect against lung cancer, however, in lung cancer chemoprevention trial study, β-carotene supplementation to smokers increased the lung cancer risk [210]. These studies raise important question/concern(s) of whether disease and cancer protection comes from a combination of phytonutrients in whole foods or isolated supplements. The other questions are about the safety, quantity, quality and the long-term effects of these dietary supplements, as these supplements can alter the body function, if given in high doses. Excessive consumption of this carotenoid can cause a deep orange discoloration of the skin, a harmless condition called lycopenodermia. Yet little is known about the effects of high doses of lycopene supplements, however, more research is required to confirm whether dietary supplements are safe and effective before marketing.

CONCLUSION

Free radicals damage has been implicated in increased risk of numerous chronic diseases, including cancer. Antioxidants reducc, prevent and repair cellular damage by neutralizing and eradicating these radicals before they interact with other molecules. Increase in free radicals production in response to normal metabolic processes, lifestyle and environmental factors induces epigenetic changes that may have a long-term effect on human health. Although research has identified numerous plant-based dietary antioxidants, only a relatively few of them have been scrutinized extensively for their long-term prevention effects. Due to the reversible nature, the epigenetic targeted therapy is one of the most promising areas for the prevention of various diseases, including cancer. Research seems to indicate that various dietary antioxidants, such as, curcumin, reseveratrol, EGCG, quercetin and genistein have an ability to modulate the altered epigenome. The extensive data summarized here in this article reviews the role of lycopene may in part be attributable to its antioxidant, antitumor and epigenetic properties in the disease risk management. Although lycopene has driven considerable interest towards preventing the progression of diseases, many aspects of mechanism of action including epigenetics still remain unclear (Figure 1).

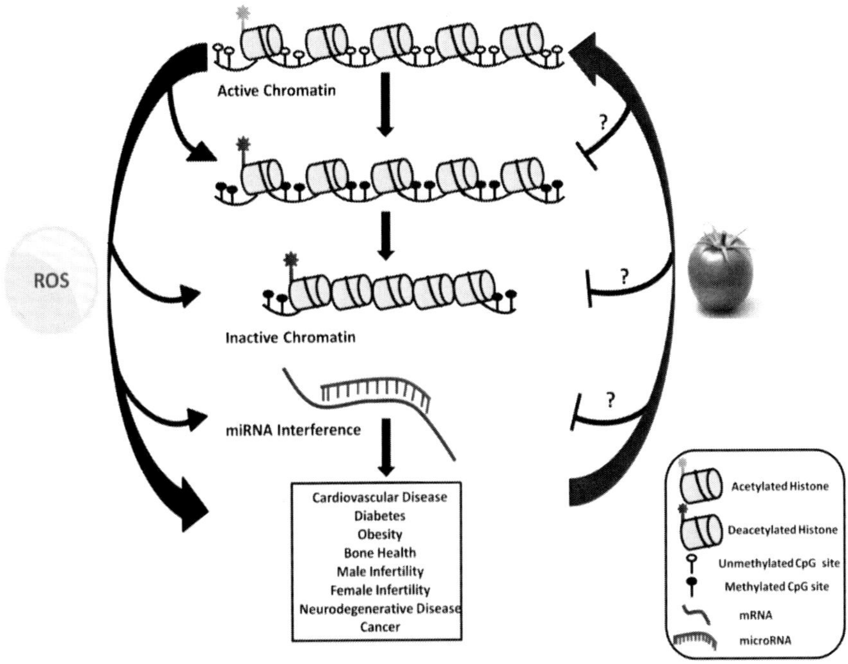

Figure 1. Schematic view showing the ROS effects on epigenetic modifications during disease development and carcinogenesis. ROS leads to the dysregulation of epigenetic control events, such as hyper/hypomethylation of CpG islands and modifies the histone marks. The effect of ROS on miRNA expression leads to translational repression. Strategy to prevent and overturn the adverse epigenetic alterations through lycopene in human diseases and cancer.

Only three studies have shed light on the epigenetic contribution of lycopene in prostate and breast cancer chemoprevention. Admittedly, there are still concerns and paucity of research in lycopene mediated epigenetic regulation in diseases. Besides this, there are many questions that remain to be answered such as:

- what dosages/concentrations of lycopene will be effective in humans?
- the challenge is also to characterize the role of dietary agents alone, or in combination with other dietary agents and assessing the long-term efficacy and safety of these agents
- whether the consumption of lycopene rich diet would result in long-term epigenetic regulation of gene expression and have strong potential in chemoprevention?
- whether the lycopene contribute to cytotoxicity at higher/lower concentrations?
- whether certain pathway or subset of genes modulated by lycopene?
- whether lycopene may be able to prevent successive generation from disease and cancer, if given in maternal diet?

There is a particular need to pay attention to long-term human intervention study to test the efficacy of lycopene in disease prevention. The exciting field of personalized nutrition and personalized dietary advice is also emerging in the near future which could be a promising strategy for disease avoidance. Further research is still required to gather additional necessary information regarding effectiveness and epigenetic changes induced by phytonutrients which will open new avenues and can result in better understanding of the impact of dietary agent in disease and cancer prevention and therapy.

Acknowledgment

We apologize to those authors whose contributions, however important, have not been cited.

References

[1] Anand, P., et al., Cancer is a preventable disease that requires major lifestyle changes. *Pharm Res,* 2008. 25(9): p. 2097-116.

[2] Abdollahi, M., et al., Pesticides and oxidative stress: a review. *Med Sci Monit,* 2004. 10(6): p. RA141-7.

[3] Hekimi, S., J. Lapointe, and Y. Wen, Taking a "good" look at free radicals in the aging process. *Trends Cell Biol,* 2011. 21(10): p. 569-76.

[4] Hasani-Ranjbar, S., B. Larijani, and M. Abdollahi, A systematic review of the potential herbal sources of future drugs effective in oxidant-related diseases. *Inflamm Allergy Drug Targets,* 2009. 8(1): p. 2-10.

[5] Halliwell, B., Free radicals, antioxidants, and human disease: curiosity, cause, or consequence? Lancet, 1994. 344(8924): p. 721-4.

[6] Ames, B.N., L.S. Gold, and W.C. Willett, The causes and prevention of cancer. *Proc Natl Acad Sci U S A,* 1995. 92(12): p. 5258-65.

[7] Harman, D., Free radical theory of aging. *Mutat Res,* 1992. 275(3-6): p. 257-66.

[8] Shigenaga, M.K., T.M. Hagen, and B.N. Ames, Oxidative damage and mitochondrial decay in aging. *Proc Natl Acad Sci U S A*, 1994. 91(23): p. 10771-8.

[9] Halliwell, B., How to characterize an antioxidant: an update. Biochem Soc Symp, 1995. 61: p. 73-101.

[10] Shi, H., N. Noguchi, and E. Niki, Comparative study on dynamics of antioxidative action of alpha-tocopheryl hydroquinone, ubiquinol, and alpha-tocopherol against lipid peroxidation. *Free Radic Biol Med*, 1999. 27(3-4): p. 334-46.

[11] Levine, M., et al., Criteria and recommendations for vitamin C intake. *JAMA,* 1999. 281(15): p. 1415-23.

[12] Steinmetz, K.A. and J.D. Potter, Vegetables, fruit, and cancer prevention: a review. *J Am Diet Assoc,* 1996. 96(10): p. 1027-39.

[13] Block, G., B. Patterson, and A. Subar, Fruit, vegetables, and cancer prevention: a review of the epidemiological evidence. *Nutr Cancer,* 1992. 18(1): p. 1-29.

[14] Gaziano, J.M., et al., A prospective study of consumption of carotenoids in fruits and vegetables and decreased cardiovascular mortality in the elderly. *Ann Epidemiol*, 1995. 5(4): p. 255-60.

[15] Giovannucci, E., A review of epidemiologic studies of tomatoes, lycopene, and prostate cancer. *Exp Biol Med* (Maywood), 2002. 227(10): p. 852-9.

[16] King-Batoon, A., J.M. Leszczynska, and C.B. Klein, Modulation of gene methylation by genistein or lycopene in breast cancer cells. *Environ Mol Mutagen,* 2008. 49(1): p. 36-45.

[17] Chalabi, N., et al., The effects of lycopene on the proliferation of human breast cells and BRCA1 and BRCA2 gene expression. *Eur J Cancer,* 2004. 40(11): p. 1768-75.

[18] Shi, J. and M. Le Maguer, Lycopene in tomatoes: chemical and physical properties affected by food processing. *Crit Rev Food Sci Nutr,* 2000. 40(1): p. 1-42.

[19] Clinton, S.K., et al., cis-trans lycopene isomers, carotenoids, and retinol in the human prostate. *Cancer Epidemiol Biomarkers Prev,* 1996. 5(10): p. 823-33.

[20] Micozzi, M.S., et al., Plasma carotenoid response to chronic intake of selected foods and beta-carotene supplements in men. *Am J Clin Nutr,* 1992. 55(6): p. 1120-5.

[21] Gerster, H., The potential role of lycopene for human health. *J Am Coll Nutr,* 1997. 16(2): p. 109-26.

[22] Stahl, W., et al., Carotenoid mixtures protect multilamellar liposomes against oxidative damage: synergistic effects of lycopene and lutein. *FEBS Lett,* 1998. 427(2): p. 305-8.

[23] Stahl, W. and H. Sies, Uptake of lycopene and its geometrical isomers is greater from heat-processed than from unprocessed tomato juice in humans. *J Nutr,* 1992. 122(11): p. 2161-6.

[24] Ip, B.C., et al., Lycopene metabolite, apo-10'-lycopenoic acid, inhibits diethylnitrosamine-initiated, high fat diet-promoted hepatic inflammation and tumorigenesis in mice. *Cancer Prev Res* (Phila), 2013. 6(12): p. 1304-16.

[25] Lian, F., et al., Apo-10'-lycopenoic acid inhibits lung cancer cell growth in vitro, and suppresses lung tumorigenesis in the A/J mouse model in vivo. *Carcinogenesis,* 2007. 28(7): p. 1567-74.

[26] Behrend, L., G. Henderson, and R.M. Zwacka, Reactive oxygen species in oncogenic transformation. *Biochem Soc Trans*, 2003. 31(Pt 6): p. 1441-4.

[27] Apel, K. and H. Hirt, Reactive oxygen species: metabolism, oxidative stress, and signal transduction. *Annu Rev Plant Biol,* 2004. 55: p. 373-99.

[28] Bergamini, C.M., et al., Oxygen, reactive oxygen species and tissue damage. *Curr Pharm Des,* 2004. 10(14): p. 1611-26.

[29] Reddy, M.B. and L. Clark, Iron, oxidative stress, and disease risk. *Nutr Rev*, 2004. 62(3): p. 120-4.

[30] Shah, A.M. and K.M. Channon, Free radicals and redox signalling in cardiovascular disease. *Heart,* 2004. 90(5): p. 486-7.

[31] Willner, C., An overview of the pathophysiology of neurodegenerative disorders. *Altern Ther Health Med,* 2004. 10(4): p. 26-34; quiz 35, 96.

[32] Pool-Zobel, B.L., et al., Consumption of vegetables reduces genetic damage in humans: first results of a human intervention trial with carotenoid-rich foods. *Carcinogenesis,* 1997. 18(9): p. 1847-50.
[33] El-Agamey, A., et al., Carotenoid radical chemistry and antioxidant/pro-oxidant properties. *Arch Biochem Biophys,* 2004. 430(1): p. 37-48.
[34] Mortensen, A., L.H. Skibsted, and T.G. Truscott, The interaction of dietary carotenoids with radical species. *Arch Biochem Biophys*, 2001. 385(1): p. 13-9.
[35] Krinsky, N.I. and K.J. Yeum, Carotenoid-radical interactions. *Biochem Biophys Res Commun,* 2003. 305(3): p. 754-60.
[36] Galano, A. and M. Francisco-Marquez, Reactions of OOH radical with beta-carotene, lycopene, and torulene: hydrogen atom transfer and adduct formation mechanisms. *J Phys Chem B*, 2009. 113(32): p. 11338-45.
[37] Bast, A., et al., Antioxidant effects of carotenoids. *Int J Vitam Nutr Res,* 1998. 68(6): p. 399-403.
[38] Renehan, A.G., et al., Insulin-like growth factor (IGF)-I, IGF binding protein-3, and cancer risk: systematic review and meta-regression analysis. *Lancet,* 2004. 363(9418): p. 1346-53.
[39] Walfisch, S., et al., Tomato lycopene extract supplementation decreases insulin-like growth factor-I levels in colon cancer patients. *Eur J Cancer Prev*, 2007. 16(4): p. 298-303.
[40] Graydon, R., et al., Effect of lycopene supplementation on insulin-like growth factor-1 and insulin-like growth factor binding protein-3: a double-blind, placebo-controlled trial. *Eur J Clin Nutr,* 2007. 61(10): p. 1196-200.
[41] Riso, P., et al., Effect of a tomato drink intervention on insulin-like growth factor (IGF)-1 serum levels in healthy subjects. *Nutr Cancer,* 2006. 55(2): p. 157-62.
[42] Vrieling, A., et al., Lycopene supplementation elevates circulating insulin-like growth factor binding protein-1 and -2 concentrations in persons at greater risk of colorectal cancer. *Am J Clin Nutr,* 2007. 86(5): p. 1456-62.
[43] Karas, M., et al., Lycopene interferes with cell cycle progression and insulin-like growth factor I signaling in mammary cancer cells. *Nutr Cancer,* 2000. 36(1): p. 101-11.
[44] Wu, W.B., et al., Inhibitory effect of lycopene on PDGF-BB-induced signalling and migration in human dermal fibroblasts: a possible target for cancer. *Biochem Soc Trans,* 2007. 35(Pt 5): p. 1377-8.

[45] Chiang, H.S., et al., Lycopene inhibits PDGF-BB-induced signaling and migration in human dermal fibroblasts through interaction with PDGF-BB. *Life Sci,* 2007. 81(21-22): p. 1509-17.

[46] Lambrechts, D., et al., VEGF pathway genetic variants as biomarkers of treatment outcome with bevacizumab: an analysis of data from the AViTA and AVOREN randomised trials. *Lancet Oncol,* 2012. 13(7): p. 724-33.

[47] Cardinault, N., et al., Lycopene but not lutein nor zeaxanthin decreases in serum and lipoproteins in age-related macular degeneration patients. *Clin Chim Acta,* 2005. 357(1): p. 34-42.

[48] Huang, C.S., J.W. Liao, and M.L. Hu, Lycopene inhibits experimental metastasis of human hepatoma SK-Hep-1 cells in athymic nude mice. *J Nutr,* 2008. 138(3): p. 538-43.

[49] Yang, C.M., et al., Growth inhibitory efficacy of lycopene and beta-carotene against androgen-independent prostate tumor cells xenografted in nude mice. *Mol Nutr Food Res*, 2011. 55(4): p. 606-12.

[50] Chen, M.L., et al., Lycopene inhibits angiogenesis both in vitro and in vivo by inhibiting MMP-2/uPA system through VEGFR2-mediated PI3K-Akt and ERK/p38 signaling pathways. *Mol Nutr Food Res,* 2012. 56(6): p. 889-99.

[51] Tang, F.Y., et al., Lycopene inhibits growth of human colon cancer cells via suppression of the Akt signaling pathway. *Mol Nutr Food Res*, 2008. 52(6): p. 646-54.

[52] Venkateswaran, V., et al., A combination of micronutrients is beneficial in reducing the incidence of prostate cancer and increasing survival in the Lady transgenic model. *Cancer Prev Res* (Phila), 2009. 2(5): p. 473-83.

[53] Venier, N.A., et al., Lycopene enhances the anti-proliferative and pro-apoptotic effects of capsaicin in prostate cancer in vitro. *Journal of Cancer Therapeutics and Research,* 2012. 1(1).

[54] Palozza, P., et al., Lycopene induces cell growth inhibition by altering mevalonate pathway and Ras signaling in cancer cell lines. *Carcinogenesis,* 2010. 31(10): p. 1813-21.

[55] Chalabi, N., et al., Expression profiling by whole-genome microarray hybridization reveals differential gene expression in breast cancer cell lines after lycopene exposure. *Biochim Biophys Acta,* 2007. 1769(2): p. 124-30.

[56] Scolastici, C., et al., Antigenotoxicity and antimutagenicity of lycopene in HepG2 cell line evaluated by the comet assay and micronucleus test. *Toxicol In Vitro,* 2008. 22(2): p. 510-4.
[57] Sanchez, I. and B.D. Dynlacht, New insights into cyclins, CDKs, and cell cycle control. *Semin Cell Dev Biol,* 2005. 16(3): p. 311-21.
[58] Kucuk, O., et al., Phase II randomized clinical trial of lycopene supplementation before radical prostatectomy. *Cancer Epidemiol Biomarkers Prev,* 2001. 10(8): p. 861-8.
[59] Ansari, M.S. and N.P. Gupta, Lycopene: a novel drug therapy in hormone refractory metastatic prostate cancer. *Urol Oncol,* 2004. 22(5): p. 415-20.
[60] Hwang, E.S. and P.E. Bowen, Cell cycle arrest and induction of apoptosis by lycopene in LNCaP human prostate cancer cells. *J Med Food,* 2004. 7(3): p. 284-9.
[61] Burgess, L.C., et al., Lycopene has limited effect on cell proliferation in only two of seven human cell lines (both cancerous and noncancerous) in an in vitro system with doses across the physiological range. *Toxicol In Vitro,* 2008. 22(5): p. 1297-300.
[62] http://nccam.nih.gov/health/antioxidants/.
[63] Hodis, H.N., et al., Serial coronary angiographic evidence that antioxidant vitamin intake reduces progression of coronary artery atherosclerosis. *JAMA,* 1995. 273(23): p. 1849-54.
[64] Morris, D.L., S.B. Kritchevsky, and C.E. Davis, Serum carotenoids and coronary heart disease. The Lipid Research Clinics Coronary Primary Prevention Trial and Follow-up Study. *JAMA,* 1994. 272(18): p. 1439-41.
[65] Rimm, E.B., et al., Vitamin E consumption and the risk of coronary heart disease in men. *N Engl J Med,* 1993. 328(20): p. 1450-6.
[66] Kohlmeier, L., et al., Lycopene and myocardial infarction risk in the EURAMIC Study. *Am J Epidemiol*, 1997. 146(8): p. 618-26.
[67] Rissanen, T., et al., Low plasma lycopene concentration is associated with increased intima-media thickness of the carotid artery wall. *Arterioscler Thromb Vasc Biol,* 2000. 20(12): p. 2677-81.
[68] Zou, Z.Y., et al., Effects of lutein and lycopene on carotid intima-media thickness in Chinese subjects with subclinical atherosclerosis: a randomised, double-blind, placebo-controlled trial. *Br J Nutr,* 2014. 111(3): p. 474-80.
[69] Baccarelli, A., et al., Ischemic heart disease and stroke in relation to blood DNA methylation. *Epidemiology,* 2010. 21(6): p. 819-28.

[70] Bruneau, B.G., Epigenetic regulation of the cardiovascular system: introduction to a review series. *Circ Res,* 2010. 107(3): p. 324-6.

[71] Shirodkar, A.V. and P.A. Marsden, Epigenetics in cardiovascular disease. *Curr Opin Cardiol,* 2011. 26(3): p. 209-15.

[72] Sharma, P., et al., Detection of altered global DNA methylation in coronary artery disease patients. *DNA Cell Biol,* 2008. 27(7): p. 357-65.

[73] Sharma, P., et al., Genome wide DNA methylation profiling for epigenetic alteration in coronary artery disease patients. *Gene,* 2014. 541(1): p. 31-40.

[74] Castro, R., et al., Homocysteine metabolism, hyperhomocysteinaemia and vascular disease: an overview. *J Inherit Metab Dis,* 2006. 29(1): p. 3-20.

[75] Libby, P., et al., Inflammation in atherosclerosis: transition from theory to practice. *Circ J,* 2010. 74(2): p. 213-20.

[76] Jia, L., et al., Methylation of FOXP3 in regulatory T cells is related to the severity of coronary artery disease. *Atherosclerosis,* 2013. 228(2): p. 346-52.

[77] Zaina, S., et al., A DNA Methylation Map of Human Atherosclerosis. *Circ Cardiovasc Genet,* 2014.

[78] Kim, J., et al., Epigenetic changes in estrogen receptor beta gene in atherosclerotic cardiovascular tissues and in-vitro vascular senescence. *Biochim Biophys Acta, 2007.* 1772(1): p. 72-80.

[79] Cash, H.L., et al., Cardiovascular disease risk factors and DNA methylation at the LINE-1 repeat region in peripheral blood from Samoan Islanders. *Epigenetics,* 2011. 6(10): p. 1257-64.

[80] Wierda, R.J., et al., Global histone H3 lysine 27 triple methylation levels are reduced in vessels with advanced atherosclerotic plaques. *Life Sci,* 2014.

[81] Stein, S. and C.M. Matter, Protective roles of SIRT1 in atherosclerosis. *Cell Cycle*, 2011. 10(4): p. 640-7.

[82] Bleijerveld, O.B., et al., Proteomics of plaques and novel sources of potential biomarkers for atherosclerosis. *Proteomics Clin Appl,* 2013. 7(7-8): p. 490-503.

[83] Findeisen, H.M., F.K. Kahles, and D. Bruemmer, Epigenetic regulation of vascular smooth muscle cell function in atherosclerosis. *Curr Atheroscler Rep,* 2013. 15(5): p. 319.

[84] Bentzon, J.F., et al., Mechanisms of plaque formation and rupture. *Circ Res,* 2014. 114(12): p. 1852-66.

[85] Eom, G.H. and H. Kook, Posttranslational modifications of histone deacetylases: implications for cardiovascular diseases. *Pharmacol Ther,* 2014. 143(2): p. 168-80.

[86] Holdt, L.M., et al., Alu elements in ANRIL non-coding RNA at chromosome 9p21 modulate atherogenic cell functions through trans-regulation of gene networks. *PLoS Genet,* 2013. 9(7): p. e1003588.

[87] Vausort, M., D.R. Wagner, and Y. Devaux, Long noncoding RNAs in patients with acute myocardial infarction. *Circ Res*, 2014. 115(7): p. 668-77.

[88] Lovren, F., et al., MicroRNA-145 targeted therapy reduces atherosclerosis. *Circulation*, 2012. 126(11 Suppl 1): p. S81-90.

[89] Horie, T., et al., MicroRNA-33 deficiency reduces the progression of atherosclerotic plaque in ApoE-/- mice. *J Am Heart Assoc,* 2012. 1(6): p. e003376.

[90] Zakkar, M., et al., Activation of Nrf2 in endothelial cells protects arteries from exhibiting a proinflammatory state. *Arterioscler Thromb Vasc Biol,* 2009. 29(11): p. 1851-7.

[91] Evans, P.C., The influence of sulforaphane on vascular health and its relevance to nutritional approaches to prevent cardiovascular disease. *EPMA J,* 2011. 2(1): p. 9-14.

[92] Juurlink, B.H., Dietary Nrf2 activators inhibit atherogenic processes. *Atherosclerosis,* 2012. 225(1): p. 29-33.

[93] Kim, J.Y., et al., Sulforaphane suppresses vascular adhesion molecule-1 expression in TNF-alpha-stimulated mouse vascular smooth muscle cells: involvement of the MAPK, NF-kappaB and AP-1 signaling pathways. *Vascul Pharmacol,* 2012. 56(3-4): p. 131-41.

[94] Miao, X., et al., Sulforaphane prevention of diabetes-induced aortic damage was associated with the up-regulation of Nrf2 and its down-stream antioxidants. *Nutr Metab* (Lond), 2012. 9(1): p. 84.

[95] Yoo, S.H., et al., Sulforaphane inhibits PDGF-induced proliferation of rat aortic vascular smooth muscle cell by up-regulation of p53 leading to G1/S cell cycle arrest. *Vascul Pharmacol,* 2013. 59(1-2): p. 44-51.

[96] Coban, D., et al., Dietary curcumin inhibits atherosclerosis by affecting the expression of genes involved in leukocyte adhesion and transendothelial migration. *Mol Nutr Food Res*, 2012. 56(8): p. 1270-81.

[97] Milenkovic, D., et al., Dietary flavanols modulate the transcription of genes associated with cardiovascular pathology without changes in their DNA methylation state. *PLoS One,* 2014. 9(4): p. e95527.

[98] Wu, L., et al., Dietary approach to attenuate oxidative stress, hypertension, and inflammation in the cardiovascular system. *Proc Natl Acad Sci U S A,* 2004. 101(18): p. 7094-9.
[99] Zhou, Z., et al., Protocatechuic aldehyde suppresses TNF-alpha-induced ICAM-1 and VCAM-1 expression in human umbilical vein endothelial cells. *Eur J Pharmacol,* 2005. 513(1-2): p. 1-8.
[100] Pirola, L., et al., Genome-wide analysis distinguishes hyperglycemia regulated epigenetic signatures of primary vascular cells. *Genome Res,* 2011. 21(10): p. 1601-15.
[101] El-Osta, A., Glycemic memory. Curr Opin Lipidol, 2012. 23(1): p. 24-9.
[102] Ling, C., et al., Epigenetic regulation of PPARGC1A in human type 2 diabetic islets and effect on insulin secretion. *Diabetologia,* 2008. 51(4): p. 615-22.
[103] Mutskov, V., et al., The human insulin gene displays transcriptionally active epigenetic marks in islet-derived mesenchymal precursor cells in the absence of insulin expression. *Stem Cells,* 2007. 25(12): p. 3223-33.
[104] Miao, F., et al., In vivo chromatin remodeling events leading to inflammatory gene transcription under diabetic conditions. *J Biol Chem,* 2004. 279(17): p. 18091-7.
[105] Reddy, M.A., et al., Role of the lysine-specific demethylase 1 in the proinflammatory phenotype of vascular smooth muscle cells of diabetic mice. *Circ Res,* 2008. 103(6): p. 615-23.
[106] Villeneuve, L.M., et al., Epigenetic histone H3 lysine 9 methylation in metabolic memory and inflammatory phenotype of vascular smooth muscle cells in diabetes. *Proc Natl Acad Sci U S A,* 2008. 105(26): p. 9047-52.
[107] Mao, Y., et al., MicroRNAs as pharmacological targets in diabetes. *Pharmacol Res,* 2013. 75: p. 37-47.
[108] Colditz, G.A., et al., Diet and risk of clinical diabetes in women. *Am J Clin Nutr,* 1992. 55(5): p. 1018-23.
[109] Brown, A.A. and F.B. Hu, Dietary modulation of endothelial function: implications for cardiovascular disease. *Am J Clin Nutr*, 2001. 73(4): p. 673-86.
[110] Ceriello, A. and R. Testa, Antioxidant anti-inflammatory treatment in type 2 diabetes. *Diabetes Care*, 2009. 32 Suppl 2: p. S232-6.
[111] Upritchard, J.E., W.H. Sutherland, and J.I. Mann, Effect of supplementation with tomato juice, vitamin E, and vitamin C on LDL oxidation and products of inflammatory activity in type 2 diabetes. *Diabetes Care,* 2000. 23(6): p. 733-8.

[112] Ozmutlu, S., S. Dede, and E. Ceylan, The effect of lycopene treatment on ACE activity in rats with experimental diabetes. *J Renin Angiotensin Aldosterone Syst,* 2012. 13(3): p. 328-33.
[113] Neyestani, T.R., et al., Physiological dose of lycopene suppressed oxidative stress and enhanced serum levels of immunoglobulin M in patients with Type 2 diabetes mellitus: a possible role in the prevention of long-term complications. *J Endocrinol Invest,* 2007. 30(10): p. 833-8.
[114] Wang, L., et al., The consumption of lycopene and tomato-based food products is not associated with the risk of type 2 diabetes in women. *J Nutr,* 2006. 136(3): p. 620-5.
[115] Akbar, S., S. Bellary, and H.R. Griffiths, Dietary antioxidant interventions in type 2 diabetes patients: a meta-analysis. *The British Journal of Diabetes & Vascular Disease,* 2011. 11(2): p. 62-68.
[116] Montonen, J., et al., Dietary antioxidant intake and risk of type 2 diabetes. *Diabetes Care*, 2004. 27(2): p. 362-6.
[117] Mody, N., et al., Oxidative stress modulates osteoblastic differentiation of vascular and bone cells. *Free Radic Biol Med,* 2001. 31(4): p. 509-19.
[118] Cortizo, A.M., et al., A possible role of oxidative stress in the vanadium-induced cytotoxicity in the MC3T3E1 osteoblast and UMR106 osteosarcoma cell lines. *Toxicology,* 2000. 147(2): p. 89-99.
[119] Kim, L., A.V. Rao, and L.G. Rao, Lycopene II--effect on osteoblasts: the carotenoid lycopene stimulates cell proliferation and alkaline phosphatase activity of SaOS-2 cells. *J Med Food*, 2003. 6(2): p. 79-86.
[120] Kakutani, Y., et al., The effects of lycopene intake and exercise on bone health in young female rats. *Journal of the International Society of Sports Nutrition,* 2011. 8(Suppl 1): p. P30-P30.
[121] Liang, H., et al., Lycopene effects on serum mineral elements and bone strength in rats. *Molecules,* 2012. 17(6): p. 7093-102.
[122] Chiba, H., et al., Hesperidin, a citrus flavonoid, inhibits bone loss and decreases serum and hepatic lipids in ovariectomized mice. *J Nutr*, 2003. 133(6): p. 1892-7.
[123] Spilmont, M., et al., Pomegranate and its derivatives can improve bone health through decreased inflammation and oxidative stress in an animal model of postmenopausal osteoporosis. *Eur J Nutr,* 2014. 53(5): p. 1155-64.
[124] Deyhim, F., et al., Citrus juice modulates bone strength in male senescent rat model of osteoporosis. *Nutrition,* 2006. 22(5): p. 559-63.

[125] Martin, R., et al., Placental calcium transporter (PMCA3) gene expression predicts intrauterine bone mineral accrual. *Bone*, 2007. 40(5): p. 1203-8.
[126] Williams, F.M. and T.D. Spector, Recent advances in the genetics of osteoporosis. *J Musculoskelet Neuronal Interact*, 2006. 6(1): p. 27-35.
[127] Sharma, J.B., et al., Effect of lycopene on pre-eclampsia and intra-uterine growth retardation in primigravidas. *Int J Gynaecol Obstet*, 2003. 81(3): p. 257-62.
[128] Maritz, G.S., M. Mutemwa, and A.X. Kayigire, Tomato juice protects the lungs of the offspring of female rats exposed to nicotine during gestation and lactation. *Pediatr Pulmonol,* 2011. 46(10): p. 976-86.
[129] Lahdetie, J., Occupation- and exposure-related studies on human sperm. *J Occup Environ Med,* 1995. 37(8): p. 922-30.
[130] Thonneau, P., et al., Occupational heat exposure and male fertility: a review. *Hum Reprod,* 1998. 13(8): p. 2122-5.
[131] Słowikowska-Hilczer, J., Xenobiotics with estrogen or antiandrogen action — disruptors of the male reproductive system. *Central European Journal of Medicine,* 2006. 1(3): p. 205-227.
[132] De Celis, R., N. Pedron-Nuevo, and A. Feria-Velasco, Toxicology of male reproduction in animals and humans. *Arch Androl,* 1996. 37(3): p. 201-18.
[133] Agarwal, A., et al., Role of oxidative stress in pathogenesis of varicocele and infertility. *Urology,* 2009. 73(3): p. 461-9.
[134] Tunc, O., H.W. Bakos, and K. Tremellen, Impact of body mass index on seminal oxidative stress. *Andrologia,* 2011. 43(2): p. 121-8.
[135] Agarwal, A., A. Hamada, and S.C. Esteves, Insight into oxidative stress in varicocele-associated male infertility: part 1. *Nat Rev Urol*, 2012. 9(12): p. 678-90.
[136] Agarwal, A., R.A. Saleh, and M.A. Bedaiwy, Role of reactive oxygen species in the pathophysiology of human reproduction. *Fertil Steril,* 2003. 79(4): p. 829-43.
[137] Gharagozloo, P. and R.J. Aitken, The role of sperm oxidative stress in male infertility and the significance of oral antioxidant therapy. *Hum Reprod,* 2011. 26(7): p. 1628-40.
[138] Marques, C.J., et al., Genomic imprinting in disruptive spermatogenesis. *Lancet,* 2004. 363(9422): p. 1700-2.
[139] Benchaib, M., et al., Quantitation by image analysis of global DNA methylation in human spermatozoa and its prognostic value in in vitro fertilization: a preliminary study. *Fertil Steril,* 2003. 80(4): p. 947-53.

[140] Flanagan, J.M., et al., Intra- and interindividual epigenetic variation in human germ cells. *Am J Hum Genet,* 2006. 79(1): p. 67-84.
[141] Faure, A.K., et al., Misregulation of histone acetylation in Sertoli cell-only syndrome and testicular cancer. *Mol Hum Reprod*, 2003. 9(12): p. 757-63.
[142] Ross, C., et al., A systematic review of the effect of oral antioxidants on male infertility. *Reprod Biomed Online*, 2010. 20(6): p. 711-23.
[143] Zini, A. and N. Al-Hathal, Antioxidant therapy in male infertility: fact or fiction? *Asian J Androl,* 2011. 13(3): p. 374-81.
[144] Showell, M.G., et al., Antioxidants for male subfertility. *Cochrane Database Syst Rev,* 2011(1): p. CD007411.
[145] Ko, E.Y. and E.S. Sabanegh, Jr., The role of over-the-counter supplements for the treatment of male infertility--fact or fiction? *J Androl,* 2012. 33(3): p. 292-308.
[146] Sarkar, P.D., T. Gupt, and A. Sahu, Comparative analysis of lycopene in oxidative stress. *J Assoc Physicians India*, 2012. 60: p. 17-9.
[147] Devaraj, S., et al., A dose-response study on the effects of purified lycopene supplementation on biomarkers of oxidative stress. *J Am Coll Nutr,* 2008. 27(2): p. 267-73.
[148] Rosato, M.P., et al., Effects of lycopene on in vitro quality and lipid peroxidation in refrigerated and cryopreserved turkey spermatozoa. *Br Poult Sci,* 2012. 53(4): p. 545-52.
[149] Gupta, N.P. and R. Kumar, Lycopene therapy in idiopathic male infertility--a preliminary report. *Int Urol Nephrol,* 2002. 34(3): p. 369-72.
[150] Mangiagalli, M.G., et al., Effect of lycopene on semen quality, fertility and native immunity of broiler breeder. *Br Poult Sci,* 2010. 51(1): p. 152-7.
[151] Aly, H.A., H.A. El-Beshbishy, and Z.M. Banjar, Mitochondrial dysfunction induced impairment of spermatogenesis in LPS-treated rats: modulatory role of lycopene. *Eur J Pharmacol,* 2012. 677(1-3): p. 31-8.
[152] Kuscu, N.K. and A. Var, Oxidative stress but not endothelial dysfunction exists in non-obese, young group of patients with polycystic ovary syndrome. *Acta Obstet Gynecol Scand,* 2009. 88(5): p. 612-7.
[153] Sugino, N., et al., Nuclear factor-kappa B is required for tumor necrosis factor-alpha-induced manganese superoxide dismutase expression in human endometrial stromal cells. *J Clin Endocrinol Metab,* 2002. 87(8): p. 3845-50.

[154] Sugino, N., et al., Withdrawal of ovarian steroids stimulates prostaglandin F2alpha production through nuclear factor-kappaB activation via oxygen radicals in human endometrial stromal cells: potential relevance to menstruation. *J Reprod Dev,* 2004. 50(2): p. 215-25.

[155] Hickey, M., et al., Long-term progestin-only contraceptives result in reduced endometrial blood flow and oxidative stress. *J Clin Endocrinol Metab,* 2006. 91(9): p. 3633-8.

[156] Swarnakar, S. and S. Paul, Curcumin arrests endometriosis by downregulation of matrix metalloproteinase-9 activity. *Indian J Biochem Biophys,* 2009. 46(1): p. 59-65.

[157] Paul, S., et al., Role of melatonin in regulating matrix metalloproteinase-9 via tissue inhibitors of metalloproteinase-1 during protection against endometriosis. *J Pineal Res*, 2008. 44(4): p. 439-49.

[158] Guney, M., et al., Regression of endometrial explants in a rat model of endometriosis treated with melatonin. *Fertil Steril,* 2008. 89(4): p. 934-42.

[159] Guney, M., et al., Effect of caffeic acid phenethyl ester on the regression of endometrial explants in an experimental rat model. *Reprod Sci,* 2007. 14(3): p. 270-9.

[160] Nahum, A., et al., Lycopene inhibition of cell cycle progression in breast and endometrial cancer cells is associated with reduction in cyclin D levels and retention of p27(Kip1) in the cyclin E-cdk2 complexes. *Oncogene,* 2001. 20(26): p. 3428-36.

[161] Levy, J., et al., Lycopene is a more potent inhibitor of human cancer cell proliferation than either alpha-carotene or beta-carotene. *Nutr Cancer,* 1995. 24(3): p. 257-66.

[162] Xu, H., et al., Anti-angiogenic effects of green tea catechin on an experimental endometriosis mouse model. *Hum Reprod*, 2009. 24(3): p. 608-18.

[163] Fambrini, M., et al., Hypermethylation of HOXA10 gene in mid-luteal endometrium from women with ovarian endometriomas. *Acta Obstet Gynecol Scand,* 2013. 92(11): p. 1331-4.

[164] Liu, M., et al., [Effect of folic acid in preventing aberrant methylation of fetal endometriosis susceptibility gene HOXA10]. *Nan Fang Yi Ke Da Xue Xue Bao,* 2013. 33(6): p. 926-9.

[165] Naqvi, H., et al., Altered genome-wide methylation in endometriosis. *Reprod Sci,* 2014. 21(10): p. 1237-43.

[166] Nasu, K., et al., Aberrant histone modification in endometriosis. *Front Biosci* (Landmark Ed), 2014. 19: p. 1202-14.
[167] Chappell, L.C., et al., Effect of antioxidants on the occurrence of pre-eclampsia in women at increased risk: a randomised trial. *Lancet,* 1999. 354(9181): p. 810-6.
[168] Beazley, D., et al., Vitamin C and E supplementation in women at high risk for preeclampsia: a double-blind, placebo-controlled trial. *Am J Obstet Gynecol,* 2005. 192(2): p. 520-1.
[169] Sibai, B.M., Prevention of preeclampsia: a big disappointment. *Am J Obstet Gynecol,* 1998. 179(5): p. 1275-8.
[170] Chouliaras, L., et al., Consistent decrease in global DNA methylation and hydroxymethylation in the hippocampus of Alzheimer's disease patients. *Neurobiol Aging,* 2013. 34(9): p. 2091-9.
[171] Bakulski, K.M., et al., Genome-wide DNA methylation differences between late-onset Alzheimer's disease and cognitively normal controls in human frontal cortex. *J Alzheimers Dis*, 2012. 29(3): p. 571-88.
[172] Zhang, K., et al., Targeted proteomics for quantification of histone acetylation in Alzheimer's disease. *Proteomics,* 2012. 12(8): p. 1261-8.
[173] Ding, H., P.J. Dolan, and G.V. Johnson, Histone deacetylase 6 interacts with the microtubule-associated protein tau. *J Neurochem*, 2008. 106(5): p. 2119-30.
[174] Graff, J., et al., An epigenetic blockade of cognitive functions in the neurodegenerating brain. *Nature,* 2012. 483(7388): p. 222-6.
[175] Coppede, F., Genetics and epigenetics of Parkinson's disease. *ScientificWorld Journal,* 2012. 2012: p. 489830.
[176] Jowaed, A., et al., Methylation regulates alpha-synuclein expression and is decreased in Parkinson's disease patients' brains. *J Neurosci*, 2010. 30(18): p. 6355-9.
[177] Matsumoto, L., et al., CpG demethylation enhances alpha-synuclein expression and affects the pathogenesis of Parkinson's disease. *PLoS One,* 2010. 5(11): p. e15522.
[178] Sano, M., et al., A controlled trial of selegiline, alpha-tocopherol, or both as treatment for Alzheimer's disease. The Alzheimer's Disease Cooperative Study. *N Engl J Med,* 1997. 336(17): p. 1216-22.
[179] Lee, S., et al., Dietary supplementation with S-adenosyl methionine delayed amyloid-beta and tau pathology in 3xTg-AD mice. *J Alzheimers Dis,* 2012. 28(2): p. 423-31.
[180] Fuso, A., et al., B-vitamin deprivation induces hyperhomocysteinemia and brain S-adenosylhomocysteine, depletes brain S-adeno

sylmethionine, and enhances PS1 and BACE expression and amyloid-beta deposition in mice. *Mol Cell Neurosci*, 2008. 37(4): p. 731-46.

[181] Fuso, A., et al., S-adenosylmethionine reduces the progress of the Alzheimer-like features induced by B-vitamin deficiency in mice. *Neurobiol Aging,* 2012. 33(7): p. 1482 e1-16.

[182] Fuso, A., et al., S-adenosylmethionine/homocysteine cycle alterations modify DNA methylation status with consequent deregulation of PS1 and BACE and beta-amyloid production. *Mol Cell Neurosci,* 2005. 28(1): p. 195-204.

[183] Masaki, K.H., et al., Association of vitamin E and C supplement use with cognitive function and dementia in elderly men. *Neurology,* 2000. 54(6): p. 1265-72.

[184] Yin, Q., et al., Lycopene attenuates insulin signaling deficits, oxidative stress, neuroinflammation, and cognitive impairment in fructose-drinking insulin resistant rats. *Neuropharmacology,* 2014. 86: p. 389-96.

[185] Polidori, M.C., et al., [Influence of vascular comorbidities on the antioxidant defense system in Alzheimer's disease]. *Dtsch Med Wochenschr,* 2012. 137(7): p. 305-8.

[186] Kumar, P., H. Kalonia, and A. Kumar, Lycopene modulates nitric oxide pathways against 3-nitropropionic acid-induced neurotoxicity. *Life Sci,* 2009. 85(19-20): p. 711-8.

[187] Kumar, P. and A. Kumar, Effect of lycopene and epigallocatechin-3-gallate against 3-nitropropionic acid induced cognitive dysfunction and glutathione depletion in rat: a novel nitric oxide mechanism. *Food Chem Toxicol,* 2009. 47(10): p. 2522-30.

[188] Liu, C.B., et al., [Effect of lycopene on oxidative stress and behavioral deficits in rotenone induced model of Parkinson's disease]. Zhongguo Ying Yong Sheng Li Xue Za Zhi, 2013. 29(4): p. 380-4.

[189] Kaur, H., S. Chauhan, and R. Sandhir, Protective effect of lycopene on oxidative stress and cognitive decline in rotenonc induced model of Parkinson's disease. *Neurochem Res,* 2011. 36(8): p. 1435-43.

[190] Salman, H., et al., Lycopene affects proliferation and apoptosis of four malignant cell lines. *Biomed Pharmacother,* 2007. 61(6): p. 366-9.

[191] Giovannucci, E., Tomatoes, tomato-based products, lycopene, and cancer: review of the epidemiologic literature. *J Natl Cancer Inst*, 1999. 91(4): p. 317-31.

[192] Giovannucci, E., et al., Intake of carotenoids and retinol in relation to risk of prostate cancer. *J Natl Cancer Inst,* 1995. 87(23): p. 1767-76.

[193] Michaud, D.S., et al., Intake of specific carotenoids and risk of lung cancer in 2 prospective US cohorts. *Am J Clin Nutr*, 2000. 72(4): p. 990-7.

[194] Levi, F., et al., Dietary intake of selected micronutrients and breast-cancer risk. *Int J Cancer,* 2001. 91(2): p. 260-3.

[195] Nkondjock, A., et al., Dietary intake of lycopene is associated with reduced pancreatic cancer risk. *J Nutr,* 2005. 135(3): p. 592-7.

[196] Chalabi, N., et al., Gene signature of breast cancer cell lines treated with lycopene. *Pharmacogenomics,* 2006. 7(5): p. 663-72.

[197] Liu, A.G. and J.W. Erdman, Jr., Lycopene and apo-10'-lycopenal do not alter DNA methylation of GSTP1 in LNCaP cells. *Biochem Biophys Res Commun,* 2011. 412(3): p. 479-82.

[198] Fu, L.J., et al., The Effects of Lycopene on the Methylation of the GSTP1 Promoter and Global Methylation in Prostatic Cancer Cell Lines PC3 and LNCaP. *Int J Endocrinol,* 2014. 2014: p. 620165.

[199] Gajendragadkar, P.R., et al., Effects of oral lycopene supplementation on vascular function in patients with cardiovascular disease and healthy volunteers: a randomised controlled trial. *PLoS One,* 2014. 9(6): p. e99070.

[200] Boileau, T.W., et al., Prostate carcinogenesis in N-methyl-N-nitrosourea (NMU)-testosterone-treated rats fed tomato powder, lycopene, or energy-restricted diets. *J Natl Cancer Inst,* 2003. 95(21): p. 1578-86.

[201] John, J.H., et al., Effects of fruit and vegetable consumption on plasma antioxidant concentrations and blood pressure: a randomised controlled trial. *Lancet,* 2002. 359(9322): p. 1969-74.

[202] Engelhard, Y.N., B. Gazer, and E. Paran, Natural antioxidants from tomato extract reduce blood pressure in patients with grade-1 hypertension: a double-blind, placebo-controlled pilot study. *Am Heart J,* 2006. 151(1): p. 100.

[203] Kim, J.Y., et al., Effects of lycopene supplementation on oxidative stress and markers of endothelial function in healthy men. *Atherosclerosis,* 2011. 215(1): p. 189-95.

[204] Ried, K., O.R. Frank, and N.P. Stocks, Dark chocolate or tomato extract for prehypertension: a randomised controlled trial. *BMC Complement* Altern Med, 2009. 9: p. 22.

[205] Paran, E., et al., The effects of natural antioxidants from tomato extract in treated but uncontrolled hypertensive patients. *Cardiovasc Drugs Ther,* 2009. 23(2): p. 145-51.

[206] Ried, K. and P. Fakler, Protective effect of lycopene on serum cholesterol and blood pressure: Meta-analyses of intervention trials. *Maturitas,* 2011. 68(4): p. 299-310.
[207] Paterson, E., et al., Supplementation with fruit and vegetable soups and beverages increases plasma carotenoid concentrations but does not alter markers of oxidative stress or cardiovascular risk factors. *J Nutr,* 2006. 136(11): p. 2849-55.
[208] Hozawa, A., et al., Circulating carotenoid concentrations and incident hypertension: the Coronary Artery Risk Development in Young Adults (CARDIA) study. *J Hypertens,* 2009. 27(2): p. 237-42.
[209] Itsiopoulos, C., et al., Can the Mediterranean diet lower HbA1c in type 2 diabetes? Results from a randomized cross-over study. *Nutr Metab Cardiovasc Dis,* 2011. 21(9): p. 740-7.
[210] Goodman, G.E., et al., The association between lung and prostate cancer risk, and serum micronutrients: results and lessons learned from beta-carotene and retinol efficacy trial. *Cancer Epidemiol Biomarkers Prev,* 2003. 12(6): p. 518-26.

In: Lycopene
Editor: Jacob R. Bailey
ISBN: 978-1-63117-927-3

Chapter 2

ANTIOXIDANT PROPERTIES OF FOOD PRODUCTS CONTAINING LYCOPENE ARE INCREASED BY THE PRESENCE OF CHLOROPHYLL

Cristina Sgherri[1]*, Usue Pérez-López[2] and Calogero Pinzino[3]

[1]Dipartimento di Scienze Agrarie, Alimentari e Agro-ambientali, University of Pisa, Pisa, Italy

[2]Departamento de Biología Vegetal y Ecología, Facultad de Ciencia y Tecnología, Universidad del País Vasco/EHU, Bilbao, Spain

[3]Istituto di Chimica dei Composti Organo Metallici (ICCOM), C.N.R., Area della Ricerca del C.N.R. di Pisa, Pisa, Italy

ABSTRACT

A diet based on the consumption of fresh fruits such as tomato has been associated with health protection and longevity, due to their nutraceutical value. The latter is influenced by the presence of antioxidant compounds such as vitamin C (ascorbic acid), vitamin E (tocopherol),

---

* Corresponding author: Cristina Sgherri, Dipartimento di Scienze Agrarie, Alimentari e Agro-ambientali, University of Pisa, Via del Borghetto 80, I-56124 Pisa, Italy; Phone: +39 050 2216609; Fax: +39 050 2210606; Email: cristina.sgherri@unipi.it.

carotenoids, flavonoids as well as phenolic acids. Indeed, these molecules are able to neutralize reactive oxygen species (ROS) and, for this reason, are worldwide recognized as beneficial for preventing human diseases among which cancer and cardiovascular pathologies.

Lycopene is a natural pigment synthesized by photosynthetic plants and represents the most abundant carotenoid in tomatoes, being present in concentrations ranging from 9-150 mg/kg depending on the variety. The increase in dietary lycopene has been associated to a decreased risk of chronic diseases and one of the possible mechanisms involved in this action is represented by its antioxidant activity.

Lycopene as well as chlorophyll are lipophilic antioxidants, contributing to the antioxidant activity of tomato berries and leaves. However, the antioxidant activity of lipid extract from leaves was shown to be 45-70-fold higher than that from berries. This is because chlorophyll behaves as a fast lipophilic antioxidant (FLA), accounting for the 38% of FLA present in the lipid extract from leaves, whereas lycopene can be classified as a slow lipophilic antioxidant (SLA) responsible for the 40% of the antioxidant activity due to the slow fraction of the lipid extract from berries. Moreover, the presence of chlorophyll in a mixture containing lycopene increased the antioxidant activity compared to lycopene alone.

The role of chlorophyll for human health, for example in reducing risk of colon cancer, is more and more recognized by scientific community and its presence in food products such as not-fully ripe berries could substantially increase their nutritional value, in combination with high concentrations of lycopene.

ABBREVIATIONS

$ABTS^{\cdot+}$	2,2'-azinobis(3-ethylbenzothiazoline-6-sulfonic acid) radical cation
AH	antioxidant
DMAPP	dimethylallyl pyrophosphate
DPPH	1,1-Diphenyl-2-Picrylhydrazyl
EPR	Electron Paramagnetic Resonance
FLA	fast lipophilic antioxidant
FS	Fremy's Salt
GGPP	geranylgeranyl pyrophosphate
GSH	glutathione
HAT	hydrogen atom transfer reaction
H_2O_2	hydrogen peroxide

IPP	isopentenyl pyrophosphate
k	decay rate constant
LDL	low-density lipoprotein
MEP	methylerythritol phosphate
MVA	mevalonic acid
$NO^{\cdot}$	nitric oxide
$^{\cdot}OH$	hydroxyl radical
$^{1}O_2$	singlet oxygen
$O_2^{\cdot -}$	superoxide radical
$ROO^{\cdot}$	peroxyl radical
R	generic radical
ROS	reactive oxygen species
SET	single electron transfer reaction
SLA	slow lipophilic antioxidant
SNP	sodium nitroprusside
SOD	superoxide dismutase
TEAC	trolox equivalent antioxidant capacity
VLDL	very-low-density lipoprotein

INTRODUCTION

Lycopene, by virtue of a large array of conjugated double bonds, and extreme hydrophobicity, exhibits many biologic features in mammalian systems mostly due to its antioxidant nature. In fact, lycopene has been reported to have important roles in preventing human diseases such as cancer and cardiovascular pathologies [1-3]. Although lycopene shows an antioxidant activity against the radical probe $DPPH^{\cdot}$ (1,1-Diphenyl-2-Picrylhydrazyl) similar to the other carotenoids [4], it represents the predominant carotenoid in human plasma. Moreover, the Mediterranean diet, characterized by a large intake of tomato fruits as well as tomato-based food products, provides about 85% of dietary lycopene.

Among dietary phytochemicals, which have been proven to play an important role in preventing diseases due to their potent antioxidant activity, there is a gap of knowledge about the health benefits of chlorophylls. Generally, they are not included in the research trials despite being the most abundant pigments (about 10-20 mg/g dry weight of leaf material) in nature [5]. However, from the few data existing, chlorophyll seems to act both as antioxidant [4-6] and as a potent inducer of mammalian phase 2 proteins that protect cells against oxidants and electrophiles [7]. Contrasting results about the antioxidant nature of chlorophyll may be explained with the different behavior shown by this pigment depending on ambient conditions, which makes the determination of its antioxidant activity particularly critical. Under controlled conditions, it has been proven that chlorophyll *b* presents the double of the antiradical activity of chlorophyll *a* [4].

Despite some evidence *in vitro*, more and more studies have been attributing to chlorophylls and chlorophyll derivatives also biological activities in the body. The most important has been related to cancer prevention and include antioxidant and antimutagenic activity, mutagen trapping, modulation of xenobiotic metabolism, and induction of apoptosis. Notwithstanding the general assumption that chlorophyll is unabsorbable by humans, there is increasing evidence that even a small absorption of chlorophyll derivatives may result in significant intracellular concentrations in target tissues, with intracellular levels similar to those of carotenoids [8]. This because the chlorophyll content of commonly consumed green vegetables typically exceeds the levels of other bioactive pigments, making it a significant contributor to the total dietary phytochemical pool.

The presence of chlorophyll in a diet rich in carotenoids, among which lycopene, increases its well-known health benefits and deserves much more attention.

1. Lycopene

1.1. Lycopene in the Plant Cell

Lycopene is one of around 600 carotenoids synthesized by many plants, bacteria, fungi, and algae [9]. Being a member of the carotenoid family, it is a lipid soluble molecule characterized by a highly unsaturated straight hydrocarbon chain with a total of 13 double bonds, 11 of which are conjugated (Figure 1).

This unique nature makes lycopene a very potent antioxidant. *In vitro* studies have shown that lycopene has a singlet oxygen quenching ability two- and ten-fold higher than β-carotene and α-tocopherol, respectively. However, lycopene lacks pro-vitamin A activity not being endowed with the terminal β-ionic rings as β-carotene is.

Because of the presence of double bonds, lycopene can exist in both the *cis* and *trans* isomeric forms. In nature, lycopene is present primarily in the all *trans* isomeric form [10]. However, it can undergo mono or poly isomerization by light, thermal energy and chemical reactions to its *cis*-isomeric forms. Chasse et al. [11] found that 5-*cys* lycopene was the most stable isomer followed by the all-*trans*, 9-*cis*, 13-*cis*, 15-*cis*, 7-*cis* and 11-*cis* isomers. Moreover, 5-*cis* lycopene was shown to have the highest antioxidant properties followed by 9-*cis*, 7-*cis*, 13-*cis*, 11-*cis* and the all-*trans* isomers.

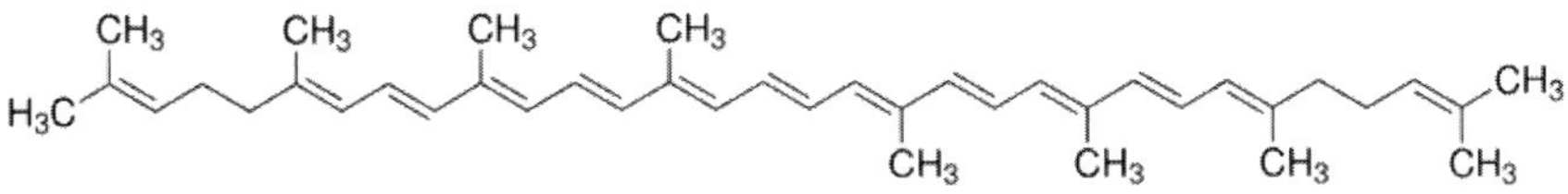

Figure 1. Chemical structure of lycopene.

Lycopene can be synthesized by the mevalonic acid (MVA) and/or by the methylerythritol phosphate (MEP) pathway. In fungi, it is only derived through the MVA cyle [12], whereas in plants it can be derived by the both aforementioned independent pathways, although it is thought that plant carotenoids synthesized in chloroplasts and chromoplasts arise primarily from the MEP pathway [13].

These pathways synthetize the five-carbon units of isopentenyl pyrophosphate (IPP) and its isomer, dimethylallyl pyrophosphate (DMAPP), needed as precursors for carotenoids [13]. Three molecules of IPP and one of DMAPP plus the presence of geranylgeranyl pyrophosphate synthase forms geranylgeranyl pyrophosphate (GGPP). From head-to-head condensation catalized by phytoene synthase, two molecules of GGPP form phytoene (Figure 2) [9, 12]. Once phytoene, molecule with nine double bonds, is formed, it is further desaturated to lycopene which contains 13 double bonds [14].

Differently from bacteria, where only one phytoene desaturase is required for converting phytoene to lycopene, formation of lycopene in plants is a multi-step process involving at least four enzymes (Figure 2). In fact, all-*trans*-lycopene molecules derived from phytoene, a 15-*cis* isomer, are produced by a phytoene desaturase and a ζ-carotene desaturase (each of them adding two symmetrical double bonds, [15]), in concert with two isomerases, a carotenoid isomerase and a 15-*cis*-ζ-carotene isomerase [16]. As the double bonds are added to the molecule, the number of conjugated double bonds increases, and the compound gains color from pale yellow (ζ-carotene) to red (lycopene).

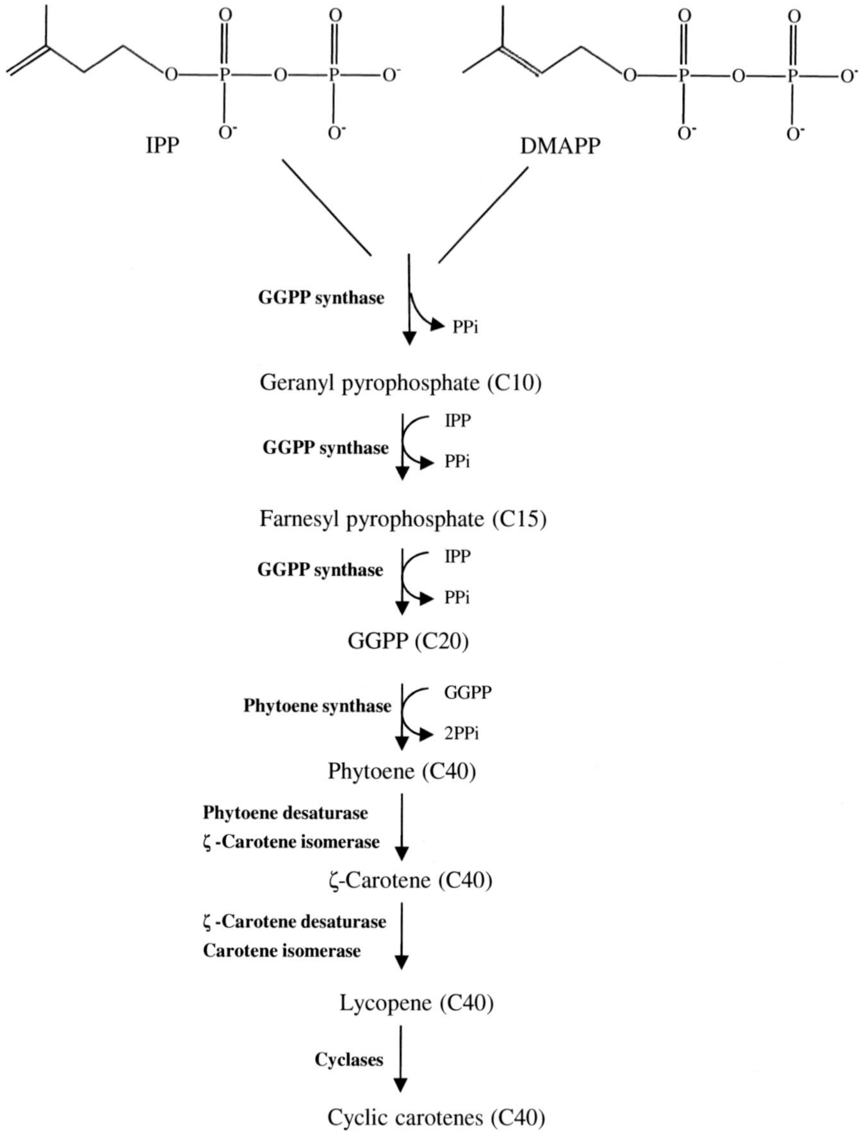

Figure 2. Carotenoid biosynthesis pathway. DMAPP, dimethylallyl pyrophosphate; GGPP, geranylgeranyl pyrophosphate; IPP, isopentenyl pyrophosphate.

Nearly all of the carotenoid biosynthetic enzymes, included those of the lycopene pathway, are located in the plastid, although all their genes are encoded by the nuclear genome [17]. Thus, carotenoid biosynthesis takes place mostly in the chloroplast envelope and, in some instances, at the thylakoid membrane [18]. In photosynthetic tissues, carotenoids accumulate in the membranes of chloroplasts [19], whereas in non-photosynthetic tissues they accumulate in chromoplasts [20], which frequently derive from fully developed chloroplasts. In all cases, chromoplasts accumulate large amounts of carotenoids in specialized lipoprotein-sequestering structures [21]. In the case of tomato (*Lycopersicon esculentum* Mill), at the breaker stage of ripening, the red color of lycopene begins to increase as the chlorophyll content decreases. In tomato fruits the induction of lycopene accumulation coincides with increased expression of upstream carotenogenic genes (phytoene synthase and phytoene desaturase) and reduced expression of genes downstream of lycopene synthesis (lycopene cyclases) [22].

1.2. Food Sources of Lycopene

In contrast to other major carotenoids consumed by humans, lycopene can be found in a limited number of foods such as some fruits, vegetables and processed foods (Table 1). Common plant sources of lycopene in the Western diet are represented by tomato, water-melon, Brazilian guava and red grapefruit [23]. Processed tomato products such as juice, ketchup, paste, sauce and soup represent good dietary sources of lycopene too. Moreover, lycopene from processed tomato products appears to be more bioavailable than that from raw tomatoes. Although tomatoes and tomato-based foods account for more than 85% of all the dietary sources of lycopene, other plant products are more and more considered possible new sources for the nutraceutical lycopene. They include autumn olive, red carrots, rose hips, bitter melon, and gac fruit (Table 1). It should be pointed out that the amount of lycopene in the different plant products is influenced by the growing site of plants, the degree of fruit ripeness, as well as by the type and conditions of processing.

Within the species considered, the content of lycopene differs also depending on cultivars. For example, in the case of tomato, concentration of lycopene in red fruits ranges from 10 to 150 mg/kg [24, 25], and cherry tomatoes generally contain significantly higher amounts of lycopene than fresh market tomato varieties [26]. Red water-melon appears to have a carotenoid synthetic pathway similar to tomato [27], and water-melon germplasm ranges in lycopene content from 30 to 120 mg/kg [28]. Red grapefruit contains 2 to 22 mg/kg lycopene, depending on variety and production environment [29]. Taking into account this genotypic variability, plant-breeding programs based on traditional and molecular methods can be used to enhance the levels of lycopene and other carotenoids in fruits.

Table 1. Lycopene contents from the main sources among fruits, vegetables and processed foods

	Lycopene (mg/kg)		Lycopene (mg/kg)
Plant Product		*Processed Foods*	
Raw tomatoes	9-150	Tomato sauce	60-180
Water-melon	23-120	Tomato paste	54-1500
Guavas	53	Tomato soup	20-80
Red grapefruits	14-34	Tomato juice	50-116
Papayas	20-53	Sun-dried tomato in oil	465
Red carrots	61	Tomatoes, cooked	37
Pitanga	73	Ketchup	42-172
Autumn olive	150-540		
Rose hips	22		
Bitter melon	411		
Gac fruit	348-1902		

The different plant products can be distinguished also for a different isomeric composition. In tomato-based foods, lycopene can be found in the *trans* form at the 79-91% and in the *cis* form at the 9-21% [30]. In red water-melon, where lycopene is the primary carotenoid, *trans* lycopene predominates reaching 92-95% of the total [31]. Although, *cis* and *trans* composition of other foods has not been well characterized yet, there are some indications that the prevalence of *cis* isomers increase the antioxidant capacity of plant products. In fact, according to Chasse et al. [11] 5-*cis* lycopene was shown to have the highest antioxidant properties followed by 9-*cis*, 7-*cis*, 13-*cis*, 11-*cis* and the all-*trans* isomers.

1.3. Effects of Environmental Conditions on Lycopene Accumulation

Besides plant-breeding programs, there are other ways to improve lycopene content in plants. These methods rely on the possibility of controlling the environmental conditions during plant growth. In fact, environmental conditions such as temperature, light, mineral availability, salinity as well as irrigation may trigger lycopene biosynthesis, being lycopene a secondary metabolite in plants. A metabolite is called secondary because of its low abundance (often less than 1% of the total carbon), and because its distribution is usually limited to dedicated cells or organs. Secondary products play a major role in the adaptation of plants to their environment, thus, they largely contribute to plant fitness by interacting with the ecosystems. It has been recognized also an antibiotic, antifungal or antiviral action, and, therefore, they are involved in the protection of plants from pathogens (phytoalexins), besides being anti-germinative or toxic for other plants (allelopathy). Moreover, secondary metabolites represent important UV absorbing compounds, thus preventing serious leaf damage from the light [32].

Because lycopene appears in chromoplasts during ripening, it has been speculated that this secondary metabolite is involved in attraction and seed consumption and dissemination by herbivores, by providing fruits with distinct red color [33]. Moreover, secondary metabolites could be retained within the ripe fruit pulp to enhance post dispersal seed germination [34]. By removing fruit pulp during consumption, frugivores would permit the germination process to be initiated.

1.3.1. Effects of Temperature

In tomato, both synthesis and degradation of lycopene are sensitive to air temperature. In particular, temperatures between 12 and 21ºC are optimal whereas values higher than 32ºC suppress its biosynthesis. For this reason, total lycopene content was found to be higher in cherry tomatoes grown in the field compared to those grown under greenhouse conditions, due to the occasional temperature build-up inside the greenhouse, which exceeded 32ºC [26]. Also Brandt et al. [35] demonstrated that when tomato fruits were subjected to a temperature higher than 30ºC lycopene content decreased significantly.

Increasing storage temperatures increases the rate of the chloroplast to chromoplast transition. Previous studies have found that temperatures below 12 and above 25 °C adversely affect the plastid conversion, causing inhibition of lycopene synthesis mainly related to the reduction of phytoene and phytofluene [36]. Moreover, storage temperatures below 12ºC may cause chilling injury, a disorder characterized by lower radical scavenging activities that increases membrane permeability, especially in green tomato fruits [37]. Thus, storage temperatures below 12ºC are responsible for the decrease in lycopene levels [38].

However, species different from tomato behave differently. In fact, in water-melon lycopene synthesis appeared not to be negatively affected by temperatures in the range 20-37ºC [39], and in red grapefruit, it was demonstrated that the synthesis was reduced when temperature range between 4 and 21ºC [40, 41].

1.3.2. Effects of Solar Radiation

At favorable temperatures (22-25ºC), McCollum [42] showed that during tomato ripening the rates of lycopene and carotene synthesis could be increased by illumination. In fact, fruits exposed to direct sunlight during their development had higher carotene levels than shaded ones. However, excessive solar radiation could affect lycopene accumulation as a consequence of radiation injury to tomato fruits associated with a general overheating on irradiated tissues [43]. Thus, lycopene production can be inhibited by excessive sunlight, the optimal conditions being represented by sufficiently high temperatures along with sufficiently dense foliage to protect the fruit from direct exposure to the sun [24].

The composition of radiation spectrum can also affect lycopene accumulation. In fact, Alba et al. [44] showed that red-light treatment stimulated lycopene accumulation in harvested mature-green fruits and that this phenomenon could be reversed by a subsequent treatment with far-red light. This light-induced lycopene accumulation was regulated by fruit-localized phytochromes and was independent from ethylene biosynthesis.

1.3.3. Effects of Mineral Uptake

Generally, lycopene accumulation is affected by nitrogen (N) amount in the soil. In tomato (cv. Moneymaker) grown in nutrient solution at three different nitrate levels the highest lycopene content was monitored under the lowest N supply [45]. In contrast, Montagu and Goh [46] found that there was a direct relation between N supply and lycopene content when tomato was grown in pots. To be optimal, N supply should be as low as possible but, at the same time, sufficient to develop a good color without reducing fruit yield.

Although potassium (K) and phosphorous (P) are needed for the synthesis of lycopene, its accumulation depends on the rate of mineral supply, besides genotype, growing conditions, and growth stage of the crop. Extra soil K was found to be necessary to avoid yellow shoulder in processing tomatoes [47], and increases from 20% to 30% in lycopene amounts were monitored in hydroponically grown tomatoes added with increasing rates of K and P supply [24]. In contrast, field experiments showed that fresh market tomatoes, watermelon, and red grapefruit had little or no enhancement in lycopene content following increased soil K [48-50].

1.3.4. Effects of Water

Both irrigation and quality of water affect lycopene accumulation.

Lycopene synthesis in response to water availability depends on crop and genotype. In watermelon irrigated at full, three-fourths or one-half of evapotranspiration rates, fruit lycopene did not significantly change [51]. In contrast, a study with different tomato cultivars (Pusa Ruby, Pusa Early Dwarf and Sioux) subjected to four irrigation regimes (40, 50, 60 and 70% depletion of available soil moisture) showed that lycopene content of fruits was reduced by moisture stress [52].

In general, being lycopene a secondary metabolite its synthesis can be triggered by environmental stresses. Thus, changes in its concentration were also monitored in red and pink tomatoes following water deficiency [53, 54] as well as in tomato grown under moderate salinity [55].

An accumulation of lycopene was also detected in other four tomato cultivars, although its magnitude was cultivar specific [56]. However, Ali and Ismail [57] showed decreases in lycopene concentrations in response of tomato to salt stress. These contradictory results could be explained by the fact that, as in case of the other secondary metabolites, changes in lycopene content depend on stress intensity [58]. Thus, lycopene synthesis can be induced under moderate salinity but inhibited as electrical conductivity exceeds the threshold value.

1.3.5. Effects of Nitric Oxide ($NO^{\cdot}$)

Nitric oxide ($NO^{\cdot}$) is a bioactive gaseous molecule involved in the signaling process within plants and plays a central role in a variety of physiological processes including germination, senescence, flowering, ripening of fruits, and response to biotic and abiotic stresses [59]. In the experiment carried out by Ali and Ismail [57], $NO^{\cdot}$ was added as sodium nitroprusside (SNP) by spraying leaves. The SNP treatment resulted in a significant increase in lycopene content, and was associated with a significant decrease in β-carotene.

Those results suggest a role of $NO^{\cdot}$ in the inhibition of the enzymes involved in β-carotene biosynthesis (e.g. β-cyclase), thus bringing to the accumulation of upstream components of the carotenoid biosynthetic pathway such as lycopene.

2. Lycopene and Human Health: Evidence of a Diet Rich in Lycopene in Contrasting the More Common Diseases

Lycopene, by virtue of its acyclic structure, large array of conjugated double bonds, and extreme hydrophobicity, exhibits many unique and distinguishing biologic features in mammalian systems. In fact, lycopene has been reported to have important roles in preventing diseases and promoting health in humans, usually associated with the reduction of the risk of cancer and cardiovascular pathologies [1-3]. However, it lacks the β-ionome ring structure and, thus, cannot form vitamin A as in the case of other carotenoids. Its biological effects in humans have therefore to be attributed to mechanisms other than vitamin A such as non-oxidative and oxidative mechanisms [60].

Lycopene represents the predominant carotenoid is human plasma and its level is affected by several biological and lifestyle factors [61]. Because the body is unable to synthesize carotenoids from endogenously produced biochemicals, human beings are totally dependent on dietary carotenoids, and tomato fruits as well as tomato-based food products provide about 85% of dietary lycopene. The remaining 15% of the total lycopene intake is obtained by the consumption of water-melon, pink grapefruit, guava, and papaya.

Owing to its hydrophobic nature, lycopene was found to be concentrated in the low-density (LDL) and very-low-density (VLDL) lipoprotein fractions of the serum [10]. This event has been related with some epidemiologic studies in which a reduced cardiovascular risk was associated with lycopene-rich diets [62-64]. Several lines of evidence have suggested that oxidatively modified LDL damages the arterial wall, and that also atherosclerosis can be attenuated by natural antioxidants (Figure 3). As reported by Fuhrman et al. [65] tomato lycopene, alone or in combinations with other natural antioxidants, can inhibit LDL oxidations. Moreover, the same group reported that dietary supplementation of tomato's lycopene (60 mg/day) for a three-month period may result in a significant reduction in plasma LDL cholesterol. This is in agreement with *in vitro* results showing that lycopene is able to suppress cholesterol synthesis and enhance LDL receptor activity in macrophages [66].

The role of lycopene in the prevention of cardiovascular disease is extensively reported in the review by Rao and Rao [67]. The strongest population-based evidence came from a multicenter case-control study (EURAMIC) that evaluated the relationship between adipose tissue antioxidative status and acute myocardial infarction [68, 69]. Subjects that included 662 cases and 717 controls were recruited from 10 different European countries. The results of this study showed a dose-response relationship between adipose tissue lycopene and the risk of myocardial infarction. Another study that compared the Lithuanian and Swedish populations showed that lower lycopene levels were associated with an increased risk and mortality from coronary heart disease [70].

Although lycopene can be found in most human tissues, it is not accumulated uniformly [71]. A preferential accumulation exists in the adrenal gland, testes, liver and prostate gland, where it is the most prominent carotenoid [10]. Differences in its distribution within the body have been related with a possible protective role in preventing degenerative diseases.

Among the numerous epidemiologic studies that correlate the fewer occurrences of cancers with diets rich in fruits and vegetables, those related with the consumption of tomato or lycopene-rich products evidenced a reduced risk for the lung, breast, stomach, and prostate cancer. Giovannucci [3] reviewed 72 epidemiological studies, including ecological, case-control, dietary and blood-specimen-based investigations of tomatoes, tomato-based products, lycopene and cancer. In 57 studies there was an inverse association between tomato intake or circulating lycopene levels and the risk of several types of cancer, and in 35 of them the association was statistically significant. None of the studies showed adverse effects of high tomato intake or high lycopene levels. The evidence for a benefit was strongest for cancers of the prostate, lung and stomach. Data were also suggestive of a benefit for cancers of the pancreas, colon and rectum, esophagus, oral cavity, breast, and cervix [3]. Because those results came from observational studies, a cause-effect relationship could not have been definitively established. However, several further studies have confirmed that increasing the intake of lycopene and its serum levels cancer risk could be reduced significantly [72-74].

With regard to the anticancer role of lycopene, prostate cancer has been the most investigated. In a study carried out by Kucuk et al. [72], 26 men with confirmed prostate cancer were supplemented with either 15 mg purified lycopene or a placebo for 3 weeks before prostatectomy. Prostate-specific antigen (PSA) levels, the biomarker of prostate dysfunction, declined by 18% in lycopene-treated subjects compared to non-supplemented men. In a smaller dosage study (8 mg lycopene per day), 40 patients with premalignant prostate neoplasia were followed for one year and then evaluated for PSA levels and prostate malignancy. The treated groups had significantly reduced PSA levels and number of prostate malignancies [75]. In a larger research project, van

Breemen et al. [76] reported that 32 men with prostate cancer and supplemented with 30 mg lycopene from tomato source for 3 weeks showed a three-fold increase in serum and prostate tissue lycopene levels. Moreover, the decline in serum PSA by 20% with lycopene treatment [77] was demonstrated to be associated with a reduction in oxidative damage to DNA (Figure 3).

Growing evidence supports more and more the protective role of lycopene also in human disorders such as osteoporosis, hypertension, and infertility. In fact, oxidative stress may contribute to the pathogenesis of the skeletal system including osteoporosis, the most prevalent metabolic bone disease [78], because oxidative stress controls the functions of both osteoclasts [79] and osteoblasts [80]. Recent studies report that antioxidants from natural sources, such as the lycopene from tomatoes, can counteract the damaging effects of oxidative stress. However, besides being an inhibitory agent of osteoclasts formation and resorption [81], lycopene seems to have also a stimulatory effect on cell proliferation [82] and on the differentiation marker alkaline phosphatase of osteoblasts [82, 83]. For the above reasons, the involvement of lycopene in bone health needs to be clarified with further investigations.

The antioxidant property of lycopene has attracted scientific research also concerning its role against hypertension and infertility. A recent study showed that lycopene supplementation at the rate of 15 mg per day for 8 weeks significantly decreased systolic blood pressures from the baseline value of 144 mmHg to 134 mmHg in mildly hypertensive subjects [84]. Another study highlighted that reduced plasmatic levels of lycopene were present in the hypertensive patients [85]. Similarly, men with antibody-mediated infertility were found to show lower semen lycopene levels than fertile controls [86]. When infertile subjects were treated with a daily dose of 8 mg lycopene for 12 months, a significant increase in serum lycopene concentration and improvements in sperm motility, sperm morphology and functional sperm concentration were observed. The lycopene treatment resulted in a 36% increase in successful pregnancies.

Due to high levels of oxygen uptake and utilization, high lipid contents and low antioxidant capacity, human brain also represents a vulnerable organ for oxidative damage. Although the role of antioxidant vitamins in neurodegenerative diseases have been extensively reported in the literature, only a small number of studies regards lycopene. During Parkinson's disease and in vascular-dementia patients, lycopene was shown to cross the blood brain barrier and to be present in the central nervous system at low concentrations. Significant reductions in the levels of lycopene in those conditions were also reported by Foy et al. [87]. Moreover, in the Austrian

Stroke Prevention study, lower amounts of serum lycopene and α-tocopherol were associated with an increased risk of microangiopathy [88], and Longnecker et al. [89] suggested a possible role of lycopene in the protection against amyotrophic lateral sclerosis disorder in humans [89].

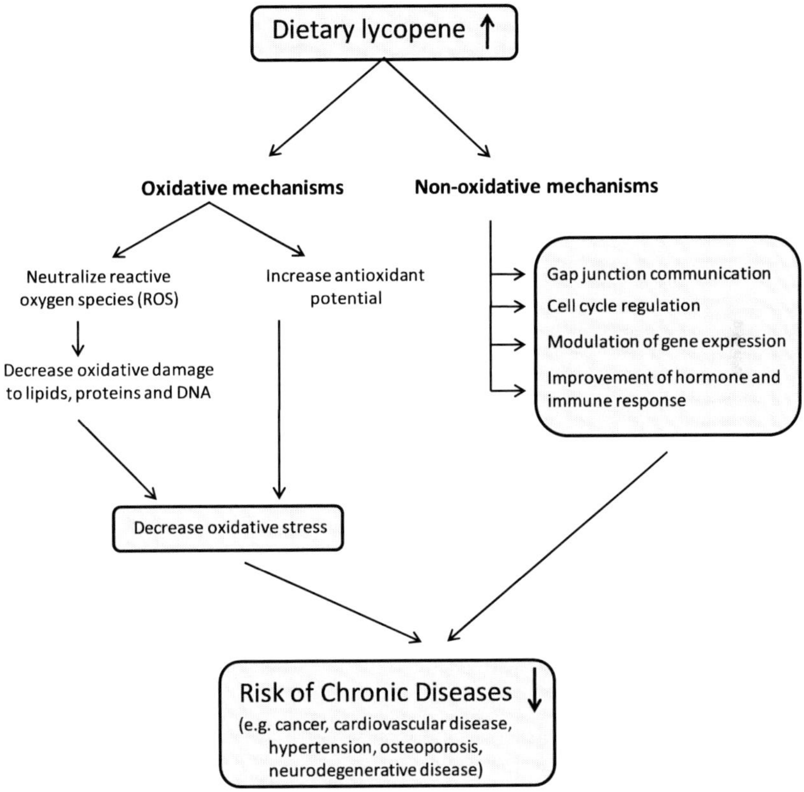

Figure 3. Oxidative and non-oxidative roles of lycopene in humans.

The scientific interest to explore innovative strategies for the prevention of human diseases underlines the common etiological and mechanistic nature of these diseases. The hypothesis that oxidation of cellular components is an initial event, eventually leading to the incidence of several diseases, brings to focus on the use of antioxidants. Examples of this hypothesis include oxidation of LDL leading to increases risk of cardiovascular disease, oxidation of DNA as an early step in the progression of cancer, and protein oxidation resulting in possible alterations in the activity of several metabolic enzymes involved in many phatologic conditions. Thus, an increase in dietary lycopene could prevent the progression of many human diseases at early stages improving the life quality (Figure 3).

Among non-oxidative roles (Figure 3), lycopene has been recognized to inhibit proliferation of several types of cancer cells, including those of breast, prostate, lung, and endometrium. The inhibitory effects of lycopene on mammary and prostate cancer cell growth were not accompanied by apoptotic (programmed) or necrotic (resulting from injury or disease) cell death, but were due to the inhibition of cell cycle progression from the G0/G1 to the S phase [90, 91]. Other roles were identified in its ability to improve intercell communications, modulate hormone and immune response, or regulate metabolism, thus lowering the risk of chronic disease (Figure 3).

Although evidence of the role of lycopene in disease prevention is suggestive, the underlying mechanisms of its activity have not been clearly understood yet. Further research is thus needed to elucidate the actual role of lycopene and to formulate dietary recommendations for healthy eating and disease prevention.

3. Role of Antioxidants

3.1. Free Radicals and Disease

Free radicals can be generated in a wide variety of chemical reactions, including the formation of plastics and the combustion of fuels, as well as in biological systems such as human body.

We need oxygen to stay alive and burn food for energy. Unfortunately, some of the oxygen molecules which are reduced during the metabolic processes give rise to reactive oxygen species (ROS) toxic for the cell. Activated oxygen can be formed by the dysfunctioning of enzymes or electron transport system, as a result of perturbations in the metabolism caused by chemical or environmental stresses. The main source of ROS is represented by electron transport chains as those localized in mitochondria or in chloroplasts [92].

In the structure of atoms and molecules, electrons usually associate in pairs, each pair moving within an atomic or molecular orbital. One electron in each pair has a spin quantum number of +1/2, the other -1/2. A free radical is any species capable of independent existence (hence the term ‘free’) that contains at least one unpaired electron, namely one electron that is alone in an orbital [93]. A dot is always used to indicate a free radical (Figure 4).

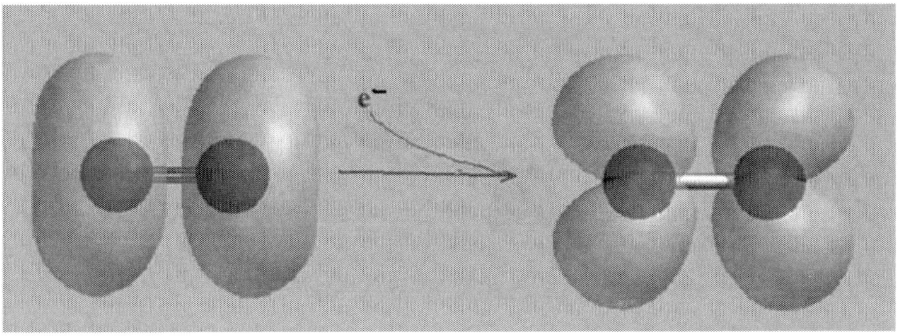

Figure 4. Superoxide anion as the product of the one-electron reduction of dioxygen (O_2).

Even if the acronym ROS is often used in the biomedical free radical literature, it represents a collective term which includes not only oxygen-centered radicals such as superoxide ($O_2^{\cdot -}$) and hydroxyl radical ($^{\cdot}OH$), but also some nonradical derivatives of oxygen such as hydrogen peroxide (H_2O_2) and singlet oxygen ($^1O_2$). However, big differences may exist in the reactivity of these species, e.g. $O_2^{\cdot -}$ and $NO^{\cdot}$ react directly with few molecules in the human body, whereas $^{\cdot}OH$ can react with anything [93], due to its high charge/surface ratio. Thus, $^{\cdot}OH$ represents the chemically most reactive species of ROS formed by successive monovalent reduction of oxygen in cell metabolism, and is primarily responsible for the cytotoxic effects of oxygen in plants, animals and micro-organisms, living in an oxygenic atmosphere. The short-lived $^{\cdot}OH$ molecule unspecifically attacks biomolecules in a diffusion-limited reaction, being able to crack polysaccharides, proteins and nucleic acids located less than a few nanometres from its site of generation. But, above all, $^{\cdot}OH$ is known for being the agent of lipid peroxidation. During this phenomenon, it is responsible for the abstracting of hydrogen from a fatty acid side-chain with the consequent formation of carbon-centered radicals and, eventually, of lipid hydroperoxides [92]. Alterations in the unsaturation of fatty acid acyl chains change fluidity of membranes and affect permeability and transport. Lipid peroxidation, thus, alters the structure and functionality of membranes damaging tissues [94]. Injury can be highly increased by the release of metal ions from storage sites. In fact, transition metals such as copper and iron stimulate conversion of the poorly reactive H_2O_2 to $^{\cdot}OH$ through the "Fenton reaction" (1), further catalyzing breakdown of lipid peroxides to peroxyl and alkoxy radicals [92].

$$H_2O_2 + Fe^{2+} \text{ (or } Cu^{+}) \longrightarrow {}^{\cdot}OH + OH^{-} + Fe^{3+} \ (Cu^{2+}) \qquad (1)$$

As a consequence, the situation of a serious imbalance between production of reactive species and antioxidant defense brings, *in vivo*, to oxidative stress.

The resulting increased oxidative damage to biomolecules may play an important role in the pathology of several human diseases (Figure 5). In fact, free radicals and other reactive species have been recognized to be implicated in the pathology of over 100 human diseases. Those that may be related to oxidative stress range from atherosclerosis, where free radicals mediate peroxidation of LDL within the vessel wall, to diseases related to chronic inflammation such as rheumatoid arthritis, inflammatory bowel disease, acute respiratory distress syndrome (ARDS), emphysema and cancers [93].

3.2. Role of Antioxidants

According to the definition of Halliwell [93] an antioxidant is any substance that, when present at low concentrations compared to an oxidizable substrate, significantly delays or prevents oxidation of that substrate. All that is due to the chemical property of acting as an electron donor. When an antioxidant encounters a free radical it freely gives an electron of its own which satisfies the free radical and stops the out-control damage. This makes, in turn, the antioxidant a free radical because it has now an electron short (Figure 6).

However, antioxidants generally possess multiple conjugated bonds or aromatic rings and are capable of delocalizing electrons, thus avoiding the abstracting of hydrogen and interrupting the free radical process.

In vivo many antioxidants come into play because different antioxidants are those needed to protect different biomolecules. This implies that there is no universal 'best' antioxidant but that each antioxidant is important relatively to which species is generated, how it is generated, where it is generated, and which is the damage target.

The antioxidant network is complex, and includes both endogenous and diet derived molecules.

In addition to enzymes, among which superoxide dismutases (SODs), which represent the first defense from ROS removing superoxide, low-molecular mass free radical scavengers exist. For example, glutathione (GSH) can scavenge various reactive species (e.g. HOCl and $ONOO^-$), both directly or through the action of glutathione peroxidases, whereas α-tocopherol (derived from the diet as vitamin E) is the most important free radical scavenger within membranes. The latter can inhibit lipid peroxidation by scavenging peroxyl radical intermediates and so halting the chain reaction. Several other antioxidants present in the diet include carotenoids, ascorbate (vitamin C) and flavonoids.

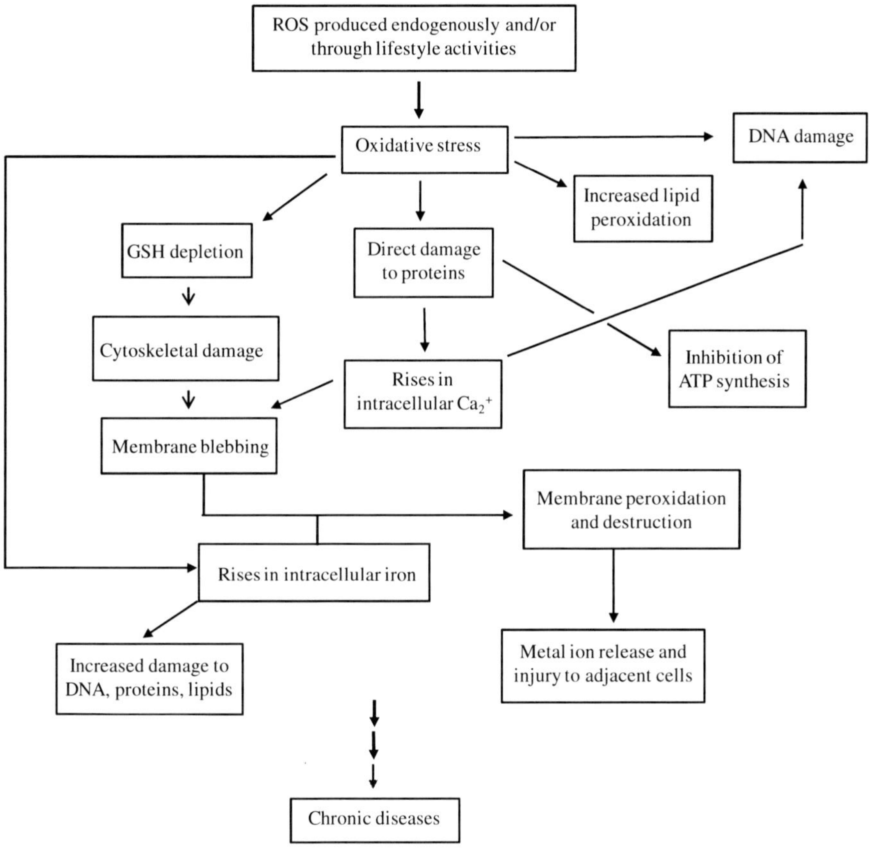

Figure 5. Oxidative stress in health and disease.

Antioxidants came to public attention in the 1990s, when scientists began to understand that free radical damage was involved in the early stages of artery-clogging atherosclerosis and could contribute to cancer, vision loss, and many other chronic conditions. Some studies showed that people with low intakes of antioxidant-rich fruits and vegetables were at greater risk for developing these chronic conditions than people who ate plenty of these fruits and vegetables. Because fruits and vegetables are rich in antioxidants, clinical trials began testing the impact of single substances, especially β-carotene and vitamin E, as weapons against heart disease, cancer, and the like.

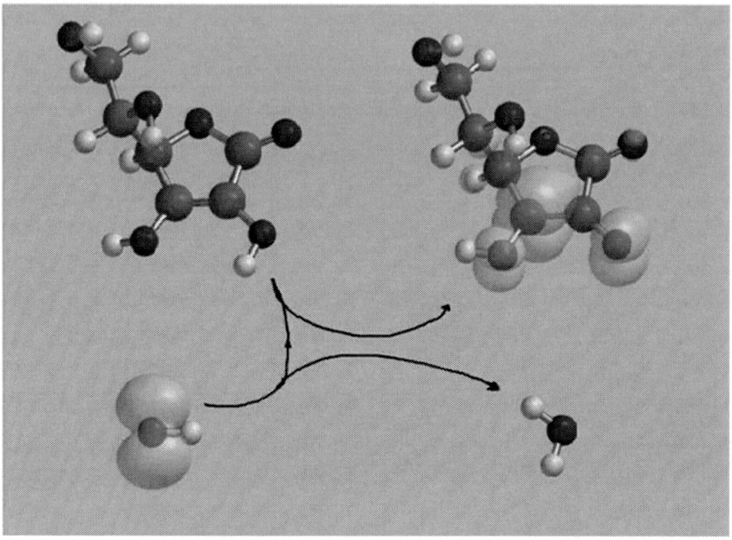

Figure 6. Free radical scavenging of antioxidants.

3.3. Antioxidant Activity of Plant Extract

The interest in traditional systems of medicine, such as Chinese and Tibetan medicines, particularly for the treatment of chronic illnesses, is more and more growing. These therapies, usually known as alternative medicine, include the use of medicinal herbs which generally refer to plant parts sometimes ground, extracted, or otherwise prepared-used for health benefits. In fact, plant derivatives can be seen as natural antioxidants helping in contrasting the progress of some pathological events whose origin resides in the generation of free radicals [95]. Moreover, they can sometimes be considered spices able to substitute synthetic antioxidants such as butylated hydroxytoluene (BHT), believed to possess carcinogenic activity, and which is employed in retarding lipid oxidation and, thus, in preventing food from deterioration [96].

The most commonly alternative therapy used is represented by dietary supplements, which include medicinal herbs and nutraceuticals. A dietary supplement may be defined as any product that contains a vitamin, mineral or amino acid, of which plants are rich.

Tomato berries constitute an example of healthy benefit plant product containing considerable amounts of vitamin C (ascorbate) and E (tocopherols), lipoic acid, phenols and minerals [97]. The interest in vitamin C, E and phenols is due to their antioxidant properties, which strongly determine their biological functions in both plant and animal metabolism. Plant products, thus, have an antioxidant activity.

Each antioxidant molecule has its own antioxidant capacity, which depends on the type and number of functional groups involved in electron donation.

Moreover, the simultaneous presence of different kind of antioxidants in a same plant extract can give rise to some synergistic effects. For example, β-carotene and vitamins E and C have a synergistic effect in cell protection, likely related to the fact that β-carotene does not only quench oxy-radicals but also repairs the α-tocopheroxyl radical [98].

Consequently, great differences can be recorded in the antioxidant capacity of different plant extracts prevalently due to the presence of different amounts of antioxidants. The extraordinary high antioxidant activity of the aqueous extract of sage, compared to that of coneflower [95], is in agreement with the observation that sage showed the highest amounts of ascorbate in comparison with other herbs and spices [99].

Antioxidant capacity also depends on the type of antioxidant which prevails in the extract and, thus, on the kind of plant extract, which can be essentially distinguished on the base of its hydrophobic/hydrophilic nature. Hydrophilic extracts contain hydrophilic antioxidants such as vitamin C, flavonoids, phenolic acids [100] whereas lipophilic ones are rich in hydrophobic antioxidants such as vitamin E, carotenoids and chlorophylls [4]. Generally, lipid extracts show an antioxidant capacity many times lower than hydrophilic extracts. For example, hydrophilic extracts from coneflower (*Echinacea angustifolia* DC.) exhibited levels of antiradical activity 5 and 7 fold-higher than lipid extracts when grown in hydroponics or soil, respectively (Figure 7).

Differences by 5-18 times were also observed in basil [100]. The reasons of that lie in the chemistry of the reaction involved, namely the biomolecule acting as reductant (type and number of functional groups involved) as well as the detection radical and the solvent used.

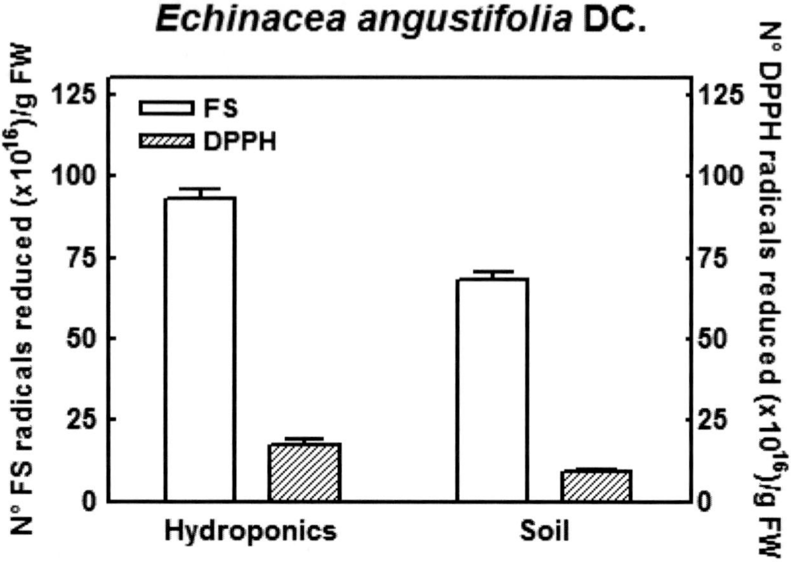

Figure 7. Antiradical activity of extracts from *Echinacea angustifolia* DC. determined by Electron Paramagnetic Resonance (EPR) following the disappearance of the free stable radicals Fremy's Salt (FS; peroxylamine disulphonate) or DPPH˙ (1,1-Diphenyl-2-Picrylhydrazyl) for hydrophilic or lipid extracts, respectively.

3.4. Measurement of Antioxidant Activity

Separating each antioxidant compound in a food mixture and studying it individually is highly costly and inefficient. Moreover, the possible synergistic interactions among the antioxidant compounds in foods make appealing to researchers to have a convenient method for a quick determination of antioxidant effectiveness in preventing diseases. However, a total antioxidant capacity assay, suitable for different food samples and which use one chemical reaction, does not exist, instead, there are numerous published methods claiming to measure total antioxidant capacity *in vitro*. These assays differ from each other in terms of substrates, probes, reaction conditions, and quantitation methods.

Basically, most of the antioxidant capacity assays can be divided into two categories: (1) hydrogen atom transfer (HAT) reaction-based assays and (2) single electron transfer (SET) reaction-based assays [101].

HAT-based methods measure the classical ability of an antioxidant (AH) to quench free radicals by hydrogen donation (2) and commonly apply a competitive reaction scheme, in which antioxidant and substrate compete for thermally generated peroxyl radicals ($ROO^{\cdot}$) through the decomposition of azo compounds:

$$ROO^{\cdot} + AH \longrightarrow ROOH + A^{\cdot} \qquad (2)$$

These assays include inhibition of induced low-density lipoprotein autoxidation, oxygen radical absorbance capacity (ORAC), total radical trapping antioxidant parameter (TRAP), inhibition of linoleic acid oxidation, and crocin bleaching assays [101].

SET-based methods detect the ability of a potential antioxidant to transfer one electron, thus reducing any compound, including metals, carbonyls, and radicals (3):

$$X^{\cdot} + AH \longrightarrow X^{-} + AH^{\cdot +} \qquad (3)$$

In general, these include assays where the measurement of the capacity of an antioxidant to reduce an oxidant is associated with change in color of the latter. The degree of color change is correlated with the sample's antioxidant concentrations. SET-based reactions include Trolox equivalent antioxidant capacity (TEAC) assay, ferric ion reducing antioxidant power (FRAP) assay, total antioxidant potential assay using Cu(II) as an oxidant, and DPPH radical scavenging capacity assay.

The ORAC assay found most of its application in the measure of the antioxidant capacity of botanical [102] and biological samples [103] whereas TEAC assay has been broadly applied in assaying food samples [104].

There is no criterion of selecting the right oxidant yet. However, Huang et al. [101] suggest that the proper molecule should be that which does not oxidize sugars, ubiquitous in food but normally not considered as antioxidants.

To comprehensively evaluate a sample's ROS scavenging capacity, more than one assays measuring individual ROS scavenging capacity are needed. In fact, radical trapping capacity directly relates to the hydrogen atom donating ability of a compound and it is not correlated to the redox potentials alone [105]. Thus, the reducing capacity of a sample is a parameter reflecting one aspect of its antioxidation property. In fact, to evaluate the radical chain-breaking antioxidant capacity the HAT-based methods result the best choice. Moreover, the effectiveness of a method is greatly influenced by the antioxidant hydrophobicity and the pH values of the assay solutions have an important effect on the reducing capacity of antioxidants [101]. At acidic conditions, the reducing capacity can be suppressed due to protonation on antioxidant compounds, whereas at basic conditions, proton dissociation would enhance a sample's reducing capacity.

DPPH$^{\cdot}$ is a well-known stable radical and a trap ("scavenger") for other radicals. Therefore, rate reduction of a chemical reaction upon DPPH$^{\cdot}$ addition can be used as indicator of the radical nature of that reaction. Because of a strong absorption band centered at about 520 nm, the DPPH radical has a deep violet color in solution, and it becomes colorless or pale yellow when neutralized. This property allows visual monitoring of the reaction, and the number of initial radicals can be counted from the change in the optical absorption at 520 nm or in the EPR signal of the DPPH$^{\cdot}$ (Figure 8). Thus, DPPH$^{\cdot}$ has two major applications in laboratory research: the first is the monitoring of chemical reactions involving radicals, most notably it is a common antioxidant assay [106], the second one regards its use as standard of the position and intensity of EPR signals. In fact, it is the most popular g-marker with an EPR splitting factor calibrated at g = 2.0036.

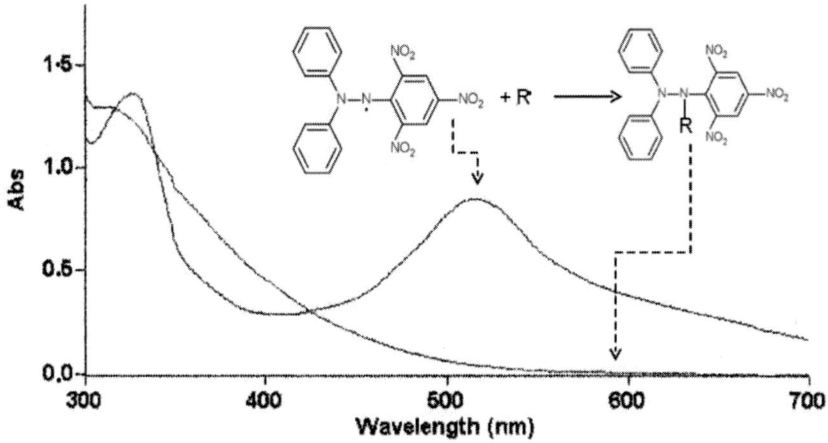

Figure 8. Spectrophotometric spectra of DPPH radical before and after trapping of another radical (R$^{\cdot}$).

Foti et al. [107] suggested that the DPPH assay involves an electron transfer reaction. In fact, in the reaction between phenols and DPPH$^{\cdot}$, the rate-determining step is represented by a fast electron transfer process from the phenoxide anions to DPPH$^{\cdot}$ whereas the hydrogen atom abstraction from the neutral aryl group by DPPH˙ is a marginal reaction occurring very slowly, especially in the presence of methanol or ethanol. Actually, the indicator radical DPPH may be neutralized either by direct reduction via electron transfers or by radical quenching via H atom transfer, being the process dependent on the composition and nature of tested antioxidants [108].

At present, none of the antioxidant capacity assays truly reflect the "total antioxidant capacity" of a particular sample [108]. In fact, total antioxidant capacity needs to reflect both lipophilic and hydrophilic capacity, and to differentiate between hydrogen atom transfer (radical quenching) and electron transfer (radical reduction). However, recently, EPR technique has been applied to the study of plant extracts allowing to add new information to the simple measure of total antioxidant capacity. Thus, kinetic parameters characterizing the hydroplilic or lipophilic behavior of antioxidants have been introduced [4, 6, 100].

EPR spectroscopy is the analytical technique that directly measures free radicals. Owing to the unpaired electron in the outer orbital, free radicals are paramagnetic species and, when present in sufficient amount, they are directly detectable by EPR spectroscopy. This technique presents many advantages over spectrophotometry as the detection at the sub-micromolar level and the possibility to analyze turbid or highly colored samples.

By using an EPR spectrometer equipped with a data acquisition system and a software package especially designed for the analysis and simulation of spectra, Sgherri et al. [4, 6] were able to monitor decay kinetics of DPPH˙ and to distinguish the kinetic behavior of different antioxidants, quantifying the type of antioxidant and determining the molar ratio of antioxidant to DPPH˙. In fact, the EPR spectrum of DPPH˙ is characterized by five lines which are narrow enough, compared to the spectrum acquisition time, to be used for registration of kinetics with many points close (Figure 9). The well-resolved quintet EPR spectrum with a_{N1} = 0.927 mT, a_{N2} = 0.846 mT at g = 2.0036 and the unimolecular decay constant of 2.06 x $10^{-6}s^{-1}$ [4, 95] make DPPH˙ a model radical to monitor kinetics lasting many hours.

One example of decay kinetics of DPPH˙ obtained in the presence of tomato leaf extract is reported in Figure 10.

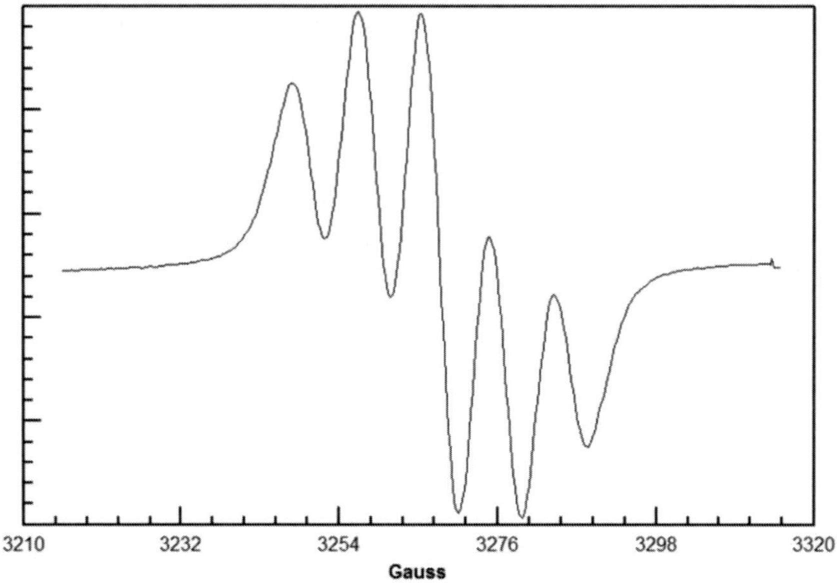

Figure 9. EPR spectrum of DPPH radical.

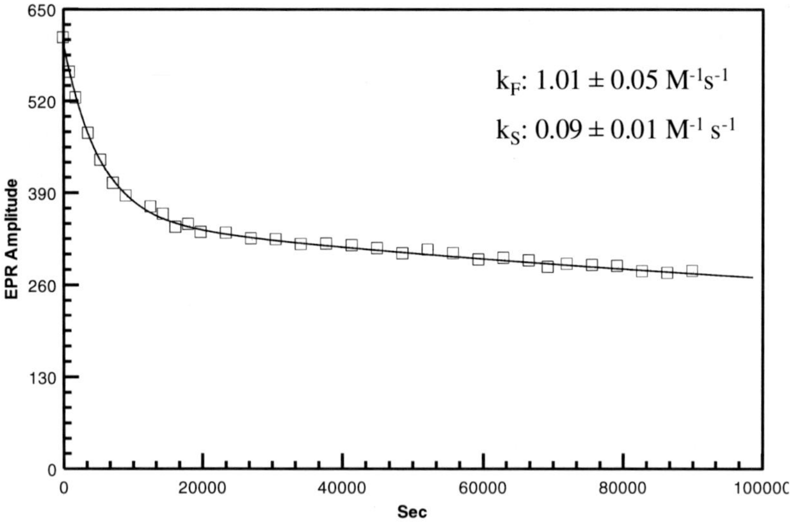

Figure 10. EPR decay kinetics of 400 μM $DPPH^{\cdot}$ following the addition of lipid extract from tomato leaves. k_F, fast-rate decay constant; k_S, slow-rate decay constant.

The application of this kinetic approach allowed to distinguish between fast and slow antioxidants in oil fractions [109] and in lipid extracts from basil [4, 100] and tomato [6]. The following equation (4), which is a special form of the pseudo-first-order kinetic model, was employed in the analyses of the experimental kinetic data [4, 6, 95, 100, 109]:

$$A = A_S \exp(-k_S' t) + A_F \exp(-k_F' t) + A_R \quad (4)$$

where A is DPPH$^{\cdot}$ molar concentration at time t, A_S and A_F are the initial DPPH$^{\cdot}$ molar concentrations that can be reduced by the slow and fast fractions, respectively, and A_R is the remaining DPPH$^{\cdot}$ concentration in the medium because of the antioxidant depletion. Eventually, k_S and k_F represent the rate constants of the slow and fast fractions, respectively, and were calculated as follows:

$$k_S = k_S'/[DPPH]_{t0}; \quad k_F = k_F'/[DPPH]_{t0}$$

Equation 4 includes the resolution of more than one differential equations of second order and its application is valid assuming that the initial molar concentration of DPPH$^{\cdot}$ ($[DPPH]_{t0}$) is greater than the molar concentration of antioxidants in the reactions. Thus, the decay kinetics of DPPH$^{\cdot}$ in the presence of lipophilic extracts is due to the contribution of two pseudo-first-order kinetics related to the presence of different antioxidants or groups of antioxidants. They differ in their velocity to scavenge the radical and the rate constants (k) can be considered as a measure of the speed of DPPH$^{\cdot}$ disappearance [4, 6, 109].

Applying this kinetic approach, k values from different plant extracts and different standard antioxidants resulted of the same order of magnitude, making them comparable [4, 6]. In this way, it could be identified the contribution of the different molecules present in plant lipid extracts to the fast and slow antioxidant activity [4], attributable to the presence of two different kinds of lipophilic antioxidants, whose action was likely distinguishable by HAT- or SET-mechanism [108].

The use of DPPH$^{\cdot}$ in ethanol mixture allows also the determination of the antioxidant power of hydrophilic extracts. Moreover, without extract, DPPH$^{\cdot}$ shows a decay constant many times lower than that of the water-soluble Fremy's salt [95], making DPPH$^{\cdot}$ more suitable for monitoring long-lasting decay kinetics. Applying this methodology, hydrophilic extracts from different cvs. of lettuce have recently been discriminated for the presence of slow-, intermediate- and fast-rate antioxidants [110].

4. Lycopene As Antioxidant

The benefits of tomato and tomato products are attributed to their antioxidant properties, especially to that of lycopene contained therein. The effectiveness of lycopene in prevention of oxidative damage to lymphocyte DNA, to cell membranes, and to HT29 cells [111] can be related to its ability to scavenge superoxide anion, hydroxyl radical, and singlet oxygen as well as to inhibit lipid peroxidation [111]. In particular, the importance of lycopene may be mainly attributed to its effective antioxidant capability against $^{\cdot}OH$, the most aggressive oxygen radical species *in vivo*.

Lycopene is a highly unsaturated acyclic isomer of β-carotene and the mechanisms of the free radical scavenging can be recognized similar to those demonstrated for β-carotene [112]. Those mechanisms can involve an electron transfer to generate the radical-cation $[\text{Lycopene}]^{\cdot+}$ (5), a hydrogen abstraction to form the carotenoid radical (6), and a radical addition to generate the adduct-radical $[R^{\cdot\cdot\cdot}\text{Lycopene}]^{\cdot}$ (7):

$$\text{Lycopene} + R \longrightarrow \text{Lycopene}]^{\cdot+} + R^{-} \text{ (electron transfer)} \quad (5)$$

$$\text{Lycopene} + R \longrightarrow \text{Lycopene}^{\cdot} + RH \text{ (allylic hydrogen abstraction)} \quad (6)$$

$$\text{Lycopene} + R \longrightarrow {}^{\cdot}[R^{\cdot\cdot\cdot}\text{Lycopene}]^{\cdot} \text{ (radical addition)} \quad (7)$$

The lycopene radical-cation and adduct-radicals are highly resonance stabilized and must therefore be relatively unreactive compared to the attacking free radical species. Thus, they undergo slow bimolecular decay to non-radical products according to reactions 8, 9:

$$2\,[\text{Lycopene}]^{\cdot+} \longrightarrow \text{product(s)} \quad (8)$$

$$2\,[R^{\cdot\cdot\cdot}\text{Lycopene}] \longrightarrow \text{product(s)} \quad (9)$$

Rates of free radical scavenging by different carotenoid antioxidants have been extensively studied by pulse radiolysis [113]. Although there were some discernible trends in carotenoid reactivity for individual radicals, rate constants varied by no greater than a factor of 2.5. The mechanism and rate of scavenging is strongly dependent on the nature of the oxidizing radical species but much less dependent on the carotenoid structure. It was shown that lycopene scavenges the sulphonylthiyl radical at a higher rate than the other carotenoids [113] and that, by the interaction with the stable radical cation $ABTS^{\cdot+}$ (2,2'-azinobis(3-ethylbenzothiazoline-6-sulfonic acid)), the ranking of carotenoids as antioxidants followed the sequence lycopene > β-carotene = β-cryptoxanthin (-3-ol) > lutein/zeaxanthin (-3,3'-diol) > echinenone (-4-one) > canthaxanthin (-4,4'-dione) = astaxanthin (-3,3'-dihydroxy-4,4'-dione) [114]. Moreover, the reactivity of lycopene towards superoxide results much more significant than that of β-carotene [113].

Application of EPR technique in $DPPH^{\cdot}$ assay becomes particularly important in the determination of antioxidant activity of carotenoids, among which lycopene. In fact, these test compounds have spectra that overlap $DPPH^{\cdot}$ at 515 nm, making interpretation of spectrophotometric results complicated.

The EPR kinetic approach was also applied to the study of the antioxidant activity of authentic standards. From the stoichiometric ratios of antioxidant to $DPPH^{\cdot}$ reported by Staško et al. [115] and Sgherri et al. [4] it can be drawn that usually the phenolic group is oxidized to carbonyl group eliminating two DPPH radicals. This is the case of vitamin E (α-, γ-, δ-tocopherol) and trolox. Moreover, antioxidants with more phenolic groups (gallic acid, quercetin) revealed a higher capacity to scavenge DPPH radicals. In contrast, other substances such as carotenoids, included lycopene, presented an antioxidant to $DPPH^{\cdot}$ ratio of 1:1 with a slow antioxidant activity [4, 6].

To reach that information, Sgherri et al. [4, 6] performed EPR experiments where the standard molar concentration was the half of that of $DPPH^{\cdot}$. In those conditions the experimental time dependencies could be fitted by the second-order kinetic model (10):

$$A = A_0(1 - r)/(1 - r\exp(-(1 - r) A_0 \mathrm{k}t)) \quad (10)$$

where A is $DPPH^{\cdot}$ molar concentration at time t; A_0 is initial $DPPH^{\cdot}$ molar concentration, k is the second-order rate constant and r is the antioxidant to $DPPH^{\cdot}$ ratio.

All carotenoids tested, namely β-Apo-8'-carotenal, lutein, β-carotene, zeaxanthin as well as lycopene, showed the same behavior against DPPH radical with slow-rate constants ranging from 0.08 (lycopene, Figure 11) to 0.13 (zeaxanthin) $M^{-1}s^{-1}$. Thus, carotenoids can generally be ascribed to slow-rate antioxidants.

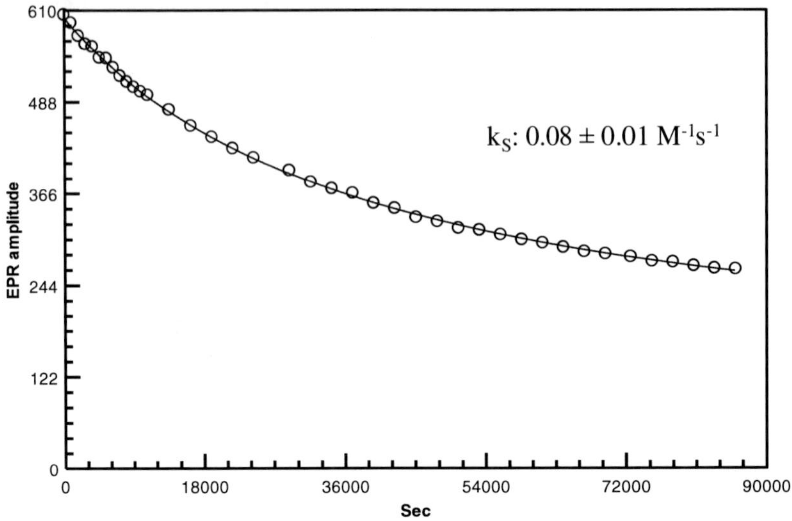

Figure 11. EPR decay kinetics of 400 μM DPPH˙ in the presence of 200 μM lycopene. k_s, slow-rate decay constant.

5. Improvement of Antioxidant Activity in the Presence of Chlorophyll

5.1. Antioxidant Activity of Chlorophyll

Chlorophyll is the natural pigment mostly widespread in the plant kingdom. It is the primary agent of the photosynthetic process occurring in the chloroplast of higher plants. Chlorophyll molecules have a hydrophobic nature and are specifically arranged in and around photosystems embedded in the thylakoid membranes. In fact, chlorophyll is a chlorin pigment structurally similar to other porphyrin pigments such as heme. At the center of the chlorin ring there is a magnesium ion. The chlorin ring can have several different side chains, usually including a long phytol. There are a few different forms that occur naturally, distinguishable for their side chains, but the most widely distributed forms in terrestrial plants are chlorophyll *a* and chlorophyll *b*.

In conditions of excess light or absence of electronic acceptors, chlorophylls are able to induce the photoreduction of oxygen giving rise to $^1O_2$ [116] and $O_2^{\cdot-}$ [117].

Differences in production and quenching of $^1O_2$ may be closely related to the molecular structure of the chlorophyll species. In fact, the yield of chlorophyll *b* is more than twice that of chlorophyll *a* and Okazaki et al. [116] supposed that replacement of side chain(s) from a methyl group (chlorophyll *a*) to an aldehyde group (chlorophyll *b*) on the R7 position on the chlorin ring induced a higher production and lower quenching of singlet oxygen. This site-specific effect of the aldehyde group suggests that one of the properties of a side chain, i.e. the electron attracting property, is the main reason for the production and quenching processes of singlet oxygen, through changes in the electronic states of molecules.

Due to these properties, chlorophyll also shows an antioxidant activity (Figure 12) [4, 6] with chlorophyll *b* displaying a rate constant of DPPH radical decay twofold higher than that of chlorophyll *a* [4]. Based on the β-carotene bleaching method, also Lanfer-Marquez et al. [5] found that *b* derivatives have higher antioxidant capacity than *a* derivatives. Decay kinetics of 400 μM DPPH$^{\cdot}$ following the addition of 200 μM chlorophyll *a* or *b* fitted well with the second-order equation (10). By the application of this kinetic model, both kinds of chlorophylls appeared to act as fast-rate antioxidants showing decay constants one order of magnitude higher than those of carotenoids (Figures 11, 12). Moreover, the stoichiometry of the reaction chlorophyll:DPPH$^{\cdot}$ was demonstrated to be 1:1 [4].

However, Endo et al. [118] observed that chlorophyll *a* had a stronger activity than BHT and chlorophyll *b*. These contrasting results can be explained, above all, with the different kinetic behavior showed by chlorophyll depending on ambient conditions, which makes determination of its antioxidant activity particularly critical. In fact, in the presence of light and reaction mixture enriched with oxygen, decay kinetics followed a pseudo-first-order reaction bringing about the complete disappearance of DPPH radical [4]. In those conditions, chlorophyll lost its antioxidant behavior characterized by the neutralization of one molecule of radical per each molecule of antioxidant. This fact can be explained with the formation of some ROS ($^1O_2$ or $O_2^{\cdot -}$) by illuminated chlorophyll which gives rise to secondary radical reactions that further neutralize DPPH$^{\cdot}$.

For both types of chlorophylls, antioxidant properties and the ability to induce photoreduction of oxygen are directly correlated *in vitro*. If this is the case *in vivo*, the use of chlorophyll *b* as the primary electron donor of the photosynthetic process might be very harmful, explaining why in nature chlorophyll *a* exerts a key role in photoreactions of photosynthetic organisms.

Catabolic derivatives of chlorophyll, such as pheophytin, chlorophyllide, pheophorbide, and pyropheophorbide, can be naturally present in plants as breakdown products but also in processed foods. For example, their presence occurs when plants degreen during plant senescence and in ripening fruits. Thus, these catabolites can be ingested preformed, but acidity in the stomach as well as microbial enzymes in the large intestine can further contribute to their production [119]. Thermal processing and/or acidification results in a perceivable discoloration of vegetable tissue from green to brown known as pheophytinization. This color loss is the result of the conversion of natural chlorophylls to Mg^{2+}-free derivatives such as pheophytins and pyropheophytins.

Notwithstanding the general assumption that chlorophyll is unabsorbable by humans, there is growing evidence about the bioavailability of its derivatives [8], making crucial the knowledge of the maintenance of the antioxidant properties following chemical modifications i.e. the loss of Mg^{2+}, phytol or acetate group. Chlorophyllins, pheophorbides, and chlorophyllides were able to reduce free radicals and protect cultured human lymphocytes against oxidative DNA damage [120]. Studying the ability to reduce $DPPH^{\cdot}$, Hsu et al. [120] showed that the relationship of free radical scavenging strengths among chlorophyll compounds is chlorophyllide > chlorophyllin > chlorophyll > pheophytin > pheophorbide. The lower ability of pyropheophorbide *a* to scavenge DPPH radical in comparison with chlorophylls was also demonstrated applying EPR technique, by the study of their relative decay kinetics (Figure 12). Because pyropheophorbide *a* showed a decay rate constant of the same order of magnitude of carotenoids, it can be classified as slow-rate antioxidant.

Although the uptake of chlorophyll derivatives seems moderately efficient, Ferruzzi and Blakeslee [8] found that their intracellular levels were similar to those of carotenoids because of the large excess of chlorophyll in purees. Thus, even a small absorption of chlorophyll derivatives may result in significant intracellular concentrations in target tissues such as the gut, proving its physiological significance.

Besides having antioxidative effects, there is evidence that chlorophyll compounds chelate Fe(II) atoms preventing the Fenton reaction [120]. In the case of chlorophylls, this would require substitution of the Fe(II) atom with the Mg one. Moreover, Arimoto-Kobayashi et al. [121] showed that Fe-chlorophyllin could bind to a planar carcinogenic molecule enhancing its degradation rate, and speculated that the Fe atom makes the Fe-chlorophyllin into an oxidative molecule that can both desmutagenically bind to planar carcinogen molecules and detoxify them by oxidation. That shed a light on the importance of meals including red meat and green vegetables because the chlorophyll from the vegetables can neutralize the iron-containing heme from the meat, reducing risk of colon cancer [6].

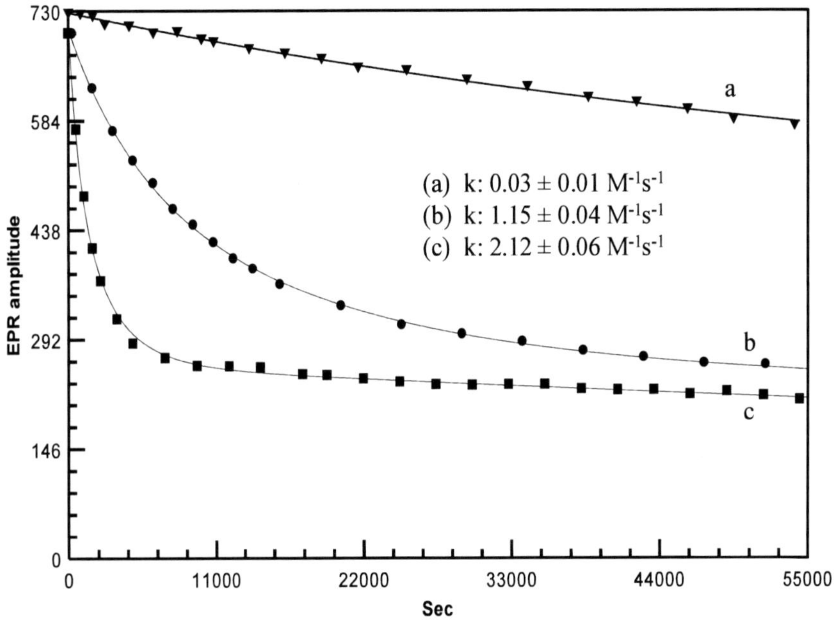

Figure 12. EPR decay kinetics of 400 μM DPPH˙ in the presence of 200 μM pyropheophorbide *a*, chlorophyll *a* (b) and chlorophyll *b* (c) monitored in the dark. k, second-order rate constant.

Notwithstanding the general thought that some antioxidants such as vitamin E play a key role in the antioxidant activity of plant extracts, they contribute minimally compared to the presence of chlorophyll. Considering the total antioxidant activity of lipophilic extracts from basil, the sum of chlorophylls *a* and *b* contributed for 40% whereas tocopherols for no more than 0.3% [4], giving further importance to a diet rich in green vegetables for human life.

5.2. Antioxidant Activity of Lycopene in the Presence of Chlorophyll

In tomato, the antioxidant activity of lipid extracts from leaves was 45- and 70-fold higher than from berries, when detected by the fitting of DPPH˙ decay kinetics and ABTS assay, respectively [6]. Moreover, tomato extracts from roots showed TEAC values approaching to zero but, by application of the pseudo-first-order kinetic model (4), antioxidant activity appeared 56-fold lower in roots than in leaves [6].

Changes in the antioxidant activity recorded in different tomato plant organs are related with the different distribution of slow lipophilic antioxidants (SLA) and fast lipophilic antioxidants (FLA), with chlorophyll (FLA) showing the majority of the antioxidant activity of leaf extracts. In contrast, lycopene, a SLA, contributed for 40% to the slow antioxidant fraction of berry lipid extracts [6]. Thus, the contemporary presence of chlorophyll and/or chlorophyll derivatives and carotenoids could represent an advantage in terms of the antioxidant activity of food products i.e. of their nutraceutical value.

To study the influence of the presence of chlorophyll (FLA) on the antioxidant activity of carotenoids (SLA), Sgherri et al. [6] applied the following second order kinetic model (11) to the fitting of the experimental time dependencies obtained with a mixture chlorophyll *b*-β-carotene:

$$A = A_F(1 - r)/(1 - r\,\exp(-(1 - r)\,A_F\,k_F t)) + A_S(1 - g)/(1 - g\,\exp(-(1 - g)\,A_S\,k_S t)) \quad (11)$$

where A is DPPH˙ molar concentration at time t; A_F and A_S are initial DPPH˙ molar concentrations which can be reduced by the FLA and SLA, respectively, k_F and k_S are the second-order rate constants of the FLA and SLA, respectively, and r and g are the ratios of chlorophyll *b*:DPPH˙ and β-carotene:DPPH˙, respectively.

The same kinetic model was applied to a mixture chlorophyll *a*-lycopene (Figure 13). In both cases, antioxidant activity of the mixture was shown to be higher compared to the carotenoid alone. Moreover, under controlled ambient conditions (dark), the stoichiometry of the reaction antioxidant–DPPH radical was maintained 1:1 both for chlorophyll *a* and lycopene, which showed decay constants similar to those obtained for each single antioxidant (Figures 11-13). Thus, under controlled conditions, the two kinds of antioxidants behaved, in a mixture, following distinct second-order reactions meaning that no synergism between chlorophyll and carotenoid molecules can be expected.

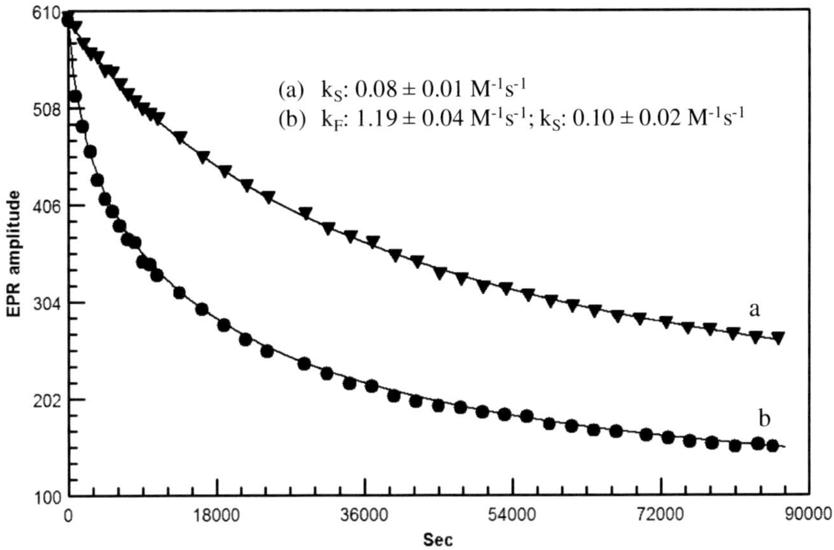

Figure 13. EPR decay kinetics of 400 μM DPPH˙ in the presence of 200 μM lycopene (a) or a mixture chlorophyll *a*-lycopene (100 μM each, b). k_S, slow-rate decay constant; k_F, fast-rate decay constant.

6. Role of Chlorophyll in Human Health

Being chlorophyll widely distributed, its content in commonly consumed green vegetables typically exceeds the levels of other bioactive pigments such as carotenoids. This relatively high concentration makes chlorophyll *a* significant contributor to the total dietary phytochemical consumption [8]. Already in 1980, Lai et al. [122] showed a relationship between antimutagenic activity and chlorophyll content for a number of common vegetable extracts. Since then numerous derivatives of chlorophyll have been identified as potential antimutagens. Endo et al. [118] first suggested that chlorophyll derivatives could be chain-breaking antioxidants by acting as effective electron donors as evidenced by its ability to reduce free radicals such as DPPH˙. Later, Sato et al. [123] identified Cu-isochlorin e_4 as a major antioxidative component of commercial grade sodium copper chlorophyllin (SCC), and Cahyana et al. [124] and Hoshina et al. [125] demonstrated a clear structural relationship within related porphyrins for the inhibition of lipid hydroperoxide formation.

Many studies have attributed to chlorophyll derivatives biological activities related to cancer prevention. They include antioxidant and antimutagenic activity, mutagen trapping, modulation of xenobiotic metabolism, and induction of apoptosis. In fact, one proposed mechanism is that the chlorophyll porphyrin ring acts as an interceptor molecule (mutagen trapping) by directly binding to other planar cyclic molecules [118, 126, 127]. Carcinogenic molecules with planar structures have been shown to be bound by natural chlorophyll which prevents them to act as carcinogens in biological systems [128]. Molecular complexes formed between chlorophyll derivatives and carcinogen molecules are physically too large to be absorbed from the gut. As a consequence, fewer carcinogen molecules reach the target organ decreasing DNA damages and the chance of tumor development. Consistent with this trapping mechanism, chlorophyllin is generally much less effective if administered after the carcinogen.

De Vogel et al. [129], in an experiment performed with rats, showed that spinach or an equimolar amount of chlorophyll inhibited damage to colonic mucosa due to dietary heme-induced luminal cytotoxicity. The combination of chlorophyll with the heme diet prevented the formation of a cytotoxic heme metabolite. They speculated that chlorophyll was able to trap heme giving rise to a hydrophobic heme-chlorophyll complex in the gut lumen, eventually inhibiting heme pro-oxidant activity [130]. This mechanism implies that intake of heme from meat should take place simultaneously to that of chlorophyll from vegetables.

Recently, chlorophyllin, a mixture of semisynthetic, water-soluble derivatives of chlorophyll used as food colorant, has also been shown to be an effective inhibitor of aflatoxin hepatocarcinogenesis in animal models by blocking the carcinogen bioavailability [131]. In fact, chlorophyllin intake at each meal led to an overall 55% reduction in urinary levels of the aflatoxin biomarker.

Limiting uptake and transport of several dietary and environmental mutagens including aflatoxin B1 and dibenzo[*a*,*l*]pyrene by chlorophyll in a Caco-2 human intestinal cells model have also demonstrated the ability of these pigments to influence bioaccessibility of carcinogens by complex formation [132, 133]. Moreover, some epidemiological evidence has linked diets rich in chlorophyll with a reduced risk of colon cancer in humans [134].

Both metal-free and metallo-chlorophyll derivatives have shown a dose-dependent inhibitory activity against benzo[*a*]pyrene induced mutagenesis, indicating the potential bioactivity of dietary chlorophyll derivatives. Thus, prophylactic interventions with chlorophyllin or diets with foods rich in chlorophylls could represent a way to prevent environmentally induced cancer.

However, chlorophyll can be also effective when not directly involved in the trapping of carcinogenic molecules, making them less powerful mutagens [127]. Other possible anti-carcinogenic mechanisms of chlorophyll are related with its antioxidant activity [5, 118, 135], being able to chelate pro-oxidant ions such as Fe(II) [136], or with its ability to stimulate cellular defenses. Furthermore, the capacity of both natural and commercial chlorophyll derivatives to act as photosensitizers have enabled their utilization as effective agents in photodynamic therapy of cancer [137, 138].

As modulators of xenobiotic detoxification systems, Singh et al. [139, 140] demonstrated in murine and suckling neonate mice that chlorophyll derivatives were able to induce glutathione-S-transferase activity, the enzyme involved in the conjugation of the reduced form of glutathione to xenobiotic substrates. On the other hand, Yun et al. [141] observed that in human liver microsomes chlorophyllin provoked a nonspecific inhibition of various CYP P450s, enzymes considered among the mostly involved in the bioactivation of the endogenous xenobiotics. Chlorophyll derivatives seem to have a role also in the apoptosis of transformed and cancerous cells. In fact, sodium copper chlorophyllin has been found to induce apoptosis in human colon cancer cells by a mechanism likely involving the activation of the caspase-8/caspase-3/poly(-ADP-ribose) polymerase pathway resulting in nuclear condensation [142]. Chan et al. [143] showed the ability of pheophorbide *a* to induce apoptosis in human hepatocellular carcinoma cells by a cytochrome c-mediated pathway involving activation of caspase-3/caspase-9. A decreased activation of mitogen protein kinases by the sodium copper chlorophyllin was observed in human breast cancer cells [144]. Although further research is required to understand the actual implications *in vivo* of natural and commercial grade chlorophyll derivatives, results obtained so far suggest that this pigment could have a great role, besides for the plant kingdom, in human health shedding a light once again on the importance of a diet rich in vegetables.

CONCLUSION

Mediterranean diet is recognized as beneficial for human health having been associated with the prevention of the most common diseases such as cancer. These health benefits are due to the presence of antioxidants such as lycopene, because tomatoes and tomato-based foods account for more than 85% of all its dietary sources. However, fruit and vegetable tissues are rich also in chlorophyll that has been proven to act as antioxidant too. In particular, green tissues showed an antioxidant activity 45-70-fold higher than tomato berries. Considering the natural abundance of chlorophyll in fruits and vegetables, the diversity of derivatives formed through food processing and preparation, and the growing use of commercial-grade chlorophyll derivatives, dietary exposure to these pigments can be significant making their intracellular levels similar to those of carotenoids. Moreover, even minimal absorption of chlorophyll derivatives seems to be physiologically significant for human health suggesting that not-fully ripe fruits present higher nutraceutical value than ripe tomatoes. All that strengthens the assertion of Demmig-Adams and Adams [145] according to which many of the compounds that protect plant cells also protect human cells, shedding a light on the importance of a diet rich in green vegetables. Indeed, the association of diets rich in phytochemicals with the prevention of cancer has intensified interest in chlorophyll as a class of plant pigments with potential chemopreventative effects.

REFERENCES

[1] Stahl, W; Sies, H. Lycopene: a biologically important carotenoid for humans? *Archives of Biochemistry and Biophysics*, 1996, 336, 1-9.

[2] Gerster, H. The potential role of lycopene for human health. *Journal of the American College of Nutrition*, 1997, 16, 109-126.

[3] Giovannucci, E. Tomatoes, tomato-based products, lycopene, and prostate cancer: review of the epidemiologic literatures. *Journal of the National Cancer Institute*, 1999, 91, 314-331.

[4] Sgherri, C; Pinzino, C; Navari-Izzo, F; Izzo R. Contribution of major lipophilic antioxidants to the antioxidant activity of basil extracts: an EPR study. *Journal of the Science of Food and Agriculture*, 2011, 91, 1128-1134.

[5] Lanfer-Marquez, UM; Barros RMC; Sinnecker, P. Antioxidant activity of chlorophylls and their derivatives. *Food Research International*, 2005, 38, 885-891.

[6] Sgherri, C; Pinzino, C; Izzo R. The role of dietary chlorophylls: an EPR study on the antioxidant activities of tomato lipid extracts. *Agrochimica*, 2011, LV, 249-260.

[7] Fahey, JW; Stephenson, KK; Dinkova-Kostova, AT; Egner, PA; Kensler, TW; Talalay, P. Chlorophyll, chlorophyllin and related tetrapyrroles are significant inducers of mammalian phase 2 cytoprotective genes. *Carcinogenesis*, 2005, 26, 1247-1255.

[8] Ferruzzi, MG; Blakeslee, J. Digestion, absorption, and cancer preventative activity of dietary chlorophyll derivatives. *Nutrition Research*, 2007, 27, 1-12.

[9] Jones, BL; Porter, JW. Biosynthesis of carotenes in higher plants. *CRC Critical Reviews in Plant Science*, 1999, 3, 295-324.

[10] Clinton, SK. Lycopene: chemistry, biology, and implications for human health and disease. *Nutrition Reviews*, 1998, 1, 35-51.

[11] Chasse, GA; Malk, ML; Deretey, E; Farkas, I; Torday, LL; Papp, JG; Sarma, DSR; Agarwal, A; Chakravarthi, S; Agarwal, S; Rao, AV. An ab initio computational study on selected lycopene isomers. *Journal of Molecular Structure (Teochem)*, 2001, 571, 27-37.

[12] Sandmann, G. Carotenoid biosynthesis and biotechnological application. *Archives of Biochemistry and Biophysics*, 2001, 385, 4-12.

[13] Botella-Pavia, P; Besumbes, O; Phillips, MA; Carretero-Paulet, L; Boronat, A; Rodriguez-Concepción, M. Regulation of carotenoid biosynthesis in plants: Evidence for a key role of hydroxymethylbutenyl diphosphate reductase in controlling the supply of plastidial isoprenoid precursors. *The Plant Journal*, 2004, 40, 188-199.

[14] Sandmann, G. Phytoene desaturase: genes, enzymes and phylogenetic aspects. *Journal of Plant Physiology*, 1994, 143, 444-447.

[15] Klein, AM; Hugueney, P; Sandmann, G; Kuntz, M. Molecular cloning and functional expression in *E. coli* of a novel plant enzyme mediating zeta-carotene desaturation. *FEBS Letters*, 1995, 372, 199-202.

[16] Hannoufa, A; Hossain, Z. Regulation of carotenoid accumulation in plants. *Biocatalysis and Agricultural Biotechnology*, 2012, 1, 198-202.

[17] Cazzonelli, CI; Cuttriss, AJ; Cossetto, SB; Pye, W; Crisp, P; Whelan, J; Finnegan, EJ; Turnbull, C; Pogson, BJ. Regulation of carotenoid composition and shoot branching in Arabidopsis by a chromatin modifying histone methyltransferase, SDG8. *The Plant Cell*, 2009, 21, 39-53.

[18] Joyard, J; Ferro, M; Masselon, C; Seigneurin-Berny, D; Salvi, D; Garin, J; Rolland, N. Chloroplast proteomics and the compartmentation of plastidial isoprenoid biosynthetic pathways. *Molecular Plant*, 2009, 2, 1154-2280.

[19] Siefermann-Harms, D. The light-harvesting and protective functions of carotenoids in photosynthetic membranes. *Physiologia Plantarum*, 1987, 69, 561-568.

[20] Camara, B; Hugueney, P; Bouvier, F; Kuntz, M; Moneger, R. Biochemistry and molecular biology of chromoplast development. *International Review of Cytology*, 1995, 163, 175-247.

[21] Vishnevetsky, M; Ovadis, M; Vainstein, A. Carotenoid sequestration in plants: the role of carotenoid-associated proteins. *Trends in Plant Science*, 1999, 4, 232-235.

[22] Bramley, PM. Regulation of carotenoid formation during tomato fruit ripening and development. *Journal of Experimental Botany*, 2002, 53, 2107-2113.

[23] Collins, JK; Perkins-Veazie, P; Roberts, W. Lycopene: From plants to humans. *HortScience*, 2006, 41, 1135-1144.

[24] Dumas, Y; Dadomo, M; Di Lucca, G; Grolier, P. Effects of environmental factors and agricultural techniques on antioxidant content of tomatoes. *Journal of the Science of Food and Agriculture*, 2003, 83, 369-382.

[25] Raffo, A; Malfa, G; Fogliano, V; Maniani, G; Quaglia, G. Seasonal variations in antioxidant components of cherry tomatoes (*Lycopersicon esculentum* cv. Naomi F1). *Journal of Food Composition and Analysis*, 2006, 19, 11-19.

[26] Kuti, JO; Konuru, HB. Effects of genotype and cultivation environment on lycopene content in red-ripe tomatoes. *Journal of the Science of Food and Agriculture*, 2005, 85, 2021-2026.

[27] Tadmor, Y; King, S; Levi, A; Davis, A; Hirschberg, J. Comparative fruit coloration in watermelon and tomato. In L. Dufose, editor. Pigment Publishing; 2004; 400-402.

[28] Perkins-Veazie, P; Collins, JK; Davis, AR; Roberts, W. Carotenoid content of 50 watermelon cultivars. *Journal of Agricultural and Food Chemistry*, 2006, 54, 2593-2597.

[29] Patil, BS; Vanamala, J; Hallman, G. Irradiation and storage influence on bioactive components and quality of early and late season "Rio Red" grapefruit. *Postharvest Biology and Technology*, 2004, 34, 53-64.

[30] Clinton, SK; Emenhiser, C; Schwartz, SJ; Bostwick, DG; Williamns, AW; Moore, BJ; Erdman JW Jr. cis-trans lycopene isomers, carotenoids, and retinol in human prostate. *Cancer Epidemiology, Biomarkers & Prevention*, 1996, 5, 823-833.

[31] Perkins-Veazie, P; Collins, JK; Pair, SD; Roberts, W. Lycopene content differs among red-fleshed watermelon cultivars. *Journal of the Science of Food and Agriculture*, 2001, 8, 1-5.
[32] Bourgaud, F; Gravot, A; Milesi, S; Gontier, E. Production of plant secondary metabolites: a historic perspective. *Plant Science*, 2001, 161, 839-851.
[33] Ronen, G; Carmel-Goren, L; Zamir, D; Hirschberg, J. An alternative pathway to b-carotene formation in plant chromoplast discovered by map-based cloning of beta and old-gold color mutations in tomato. *Proceedings of the National Academy of Sciences*, 2000, 97, 11102-11107.
[34] Cipollini, ML. Secondary metabolites of vertebrate-dispersed fruits: evidence for adaptive functions. *Revista Chilena de Historia Natural*, 2000, 73, 421-433.
[35] Brandt, S; Pék, Z; Barna, E; Lugasi, A; Helyes, L. Lycopene content and colour of ripening tomatoes as affected by environmental conditions. *Journal of Science and Food Agriculture*, 2006, 86, 568-572.
[36] Gautier, H; Diakou-Verdin, V; Bénard, C; Reich, M; Buret, M; Bourgaud, F; Poëssel, JL; Caris-Veyrat, C; Génard, M. How does tomato quality (sugar, acid, and nutritional quality) vary with ripening stage, temperature, and irradiance? *Journal of Agricultural and Food Chemistry*, 2008, 56, 1241-1250.
[37] Polenta, G; Lucangeli, C; Budde, C; González, CB; Murray, R. Heat and anaerobic treatments affected physiological and biochemical parameters in tomato fruits. LWT - *Food Science and Technology*, 2006, 39, 27-34.
[38] Farneti, B; Schouten, RE; Woltering, EJ. Low temperature-induced lycopene degradation in red ripe tomato evaluated by remittance spectroscopy. *Postharvest Biology and Technology*, 2012, 73, 22-27.
[39] Vogele, AC. Effect of environmental factors upon the color of the tomato and the watermelon. *Plant Physiology*, 1937, 12, 929-955.
[40] Purcell, AE; Young, RH; Schultz, EF; Meredith, FI. The effect of artificial climate on the interal fruit color of Redblush grapefruit. *Proceedings of the American Society for Horticultural Science*, 1968, 92, 170-178.
[41] Meredith, FI; Young, RH. Changes in lycopene and carotene content of "Redblush" grapefruit exposed to high temperatures. *HortScience*, 1971, 6, 233-234.
[42] McCollum, JP. Effects of light on the formation of carotenoids in tomato fruits. *Journal of Food Science*, 1954, 19, 182-189.

[43] Hamauzu, Y; Chachin, K; Ueda, Y. Effect of postharvest temperature on the converstion of 14C-mevalonic acid to carotenes in tomato fruit. *Journal of the Japanese Society for Horticultural Science*, 1998, 67, 549-555.

[44] Alba, R; Cordonnier-Pratt, MM; Pratt, LH. Fruit-localized phytochromes regulate lycopene accumulation independently of ethylene production in tomato. *Plant Physiology*, 2000, 123, 363-370.

[45] Aziz, ABA. Seasonal changes in the physical and chemical composition of tomato fruits as affected by nitrogen levels. *Mededelingen Landbouwhogeschool Wageningen*, Nederland, 1968, 68, 7.

[46] Montagu, KD; Goh, KM. Effects of forms and rates of organic and inorganic nitrogen fertilizers on the yield and some quality indices of tomatoes (*Lycopersicon esculentum* Miller). *New Zealand Journal of Crop and Horticultural Science*, 1990, 18, 31-37.

[47] Hartz, TK; Miyao, G; Mullen, RJ; Cahn, MD; Valencia, J; Brittan, KL. Potassium requirements for maximum yield and fruit quality of processing tomato. *Journal of the American Society for Horticultural Science*, 1999, 124, 199-204.

[48] Fontes, PR; Sampaio, RA; Finger, FL. Fruit size, mineral composition and quality of trickle-irrigated tomatoes as affected by potassium rates. *Pesquisa Agropecuária Brasileira*, 2000, 35, 21-25.

[49] Patil, BS; Alva, A. Functional components in citrus: Alteration by mineral elements. *Proceedings of the Symposium on Fertilizing Crops for Functional Food*, 2002, 7-1-7-4.

[50] Perkins-Veazie, P; Roberts, W. Can potassium application affect the mineral and antioxidand content of horticultural crops? *Proceedings of the Symposium on Fertilizing Crops for Functional Food*, 2002, 2-1-2-6.

[51] Leskovar, DI; Bang, H; Crosby, KM; Maness, N; Franco, JA; Perkins-Veazie, P. Lycopene, carbohydrates, ascorbic acid and yield components of diploid and triploid watermelon cultivars are affected by deficit irrigation. *The Journal of Horticultural Science and Biotechnology*, 2004, 79, 75-81.

[52] Naphade, AS. Effect of water regime on the quality of tomato. *Maharashtra Journal of Horticulture*, 1993, 7, 55-60.

[53] Matsuzoe, N; Zushi, K; Johjima, T. Effect of soil water deficit on coloring and carotene formation in fruits of red, pink and yellow type cherry tomatoes. *Journal of the Japanese Society for Horticultural Science*, 1998, 67, 600-606.

[54] Zushi, K; Matsuzoe, N. Effect of soil water deficit in vitamin C, sugar, organic acid, amino acid and carotene contents of large-fruited tomatoes. *Journal of Japanese Society for Horticultural Science*, 1998, 67, 927-933.

[55] Wu, M; Buck, JS; Kubota, C. Effecs of nutrient solution EC, plant microclimate and cultivars on fruit quality and yield of hydroponic tomatoes (*Lycopersicon esculentum*). *Acta Horticulturae*, 2004. 659; 541-547.

[56] Kubota, C; Thomson. CA; Wu, M; Javanmardi, J. Controlled environments for production of value-added food crops with phytochemical concentrations: lycopene in tomato as an example. *HortScience*, 2006, 41, 522-525.

[57] Ali, HEM; Ismail, GSM. Tomato fruit quality as influenced by salinity and nitric oxide. *Turkish Journal of Botany*, 2014, 38, 122-129.

[58] Sgherri, CLM; Navari-Izzo, F. Sunflower seedlings subjected to increasing water deficit stress: oxidative stress and defence mechanisms. *Physiologia Plantarum*, 1995, 93, 25-30.

[59] Zheng, C; Jiang, D; Dai, T; Jing, Q; Cao, W. Effects nitroprusside, a nitric oxide donor, on carbon and nitrogen metabolism and the activity of the antioxidation system in wheat seedlings under salt stress. *Acta Ecologica Sinica*, 2010, 30, 1174-1183.

[60] Agarwal, S; Rao, AV. Tomato lycopene and its role in human health and chronic diseases. *Canadian Medical Association Journal*, 2000, 163, 739-744.

[61] Rao, AV; Agarwal, S. Role of lycopene as antioxidant carotenoid in the prevention of chronic diseases: a review. *Nutrition Research*, 1999, 19, 305-323.

[62] Rissanen, TH; Voutilainen, S; Nyyssonen, K; Lakka, TA; Sivenius, J; Salonen, R; Kaplan, GA; Salonen, JT. Low serum lycopene concentration is associated with an excess incidence of acute coronary events and stroke: The Kuopio Ischaemic Heart Disease Risk Factor Study. *British Journal of Nutrition*, 2001, 85, 749-754.

[63] Gianetti, JR; Pedrinelli, R; Petrucci, R; Lazzerini, G; De Caterina, M; Bellomo, G; De Caterina R. Inverse association between carotid intima-media thickness and the antioxidant lycopene in atherosclerosis. *American Heart Journal*, 2002, 143, 467-474.

[64] Sesso, HD; Liu, S; Gaziano, JM; Buring, JE. Dietary lycopene, tomato-based food products and cardiovascular disease in women. *Journal of Nutrition*, 2003, 133, 2336-2341.

[65] Furhman, B; Ben-Yaish, L; Attias, J; Hayek, T; Aviram, M. Tomato lycopene and beta-carotene inhibit low density lipoprotein oxidation and this effect depends on lipoprotein vitamin E content. *Nutrition, Metabolism & Cardiovascular Diseases*, 1997, 7, 433-443.

[66] Furhman, B; Elis, A; Aviram, M. Hypocholesterolemic effect of lycopene and beta-carotene is related to suppression of cholesterol synthesis and augmentation of LDL receptor activity in macrophages. *Biochemical and Biophysical Research Communications*, 1997, 233, 658-662.

[67] Rao, AV; Rao, LG. Carotenoids and human health. *Pharmacological Research*, 2007, 55, 207-216.

[68] Kohlmeier, L; Hastings, SB. Epidemiologic evidence of a role of carotenoids in cardiovascular disease prevention. *American Journal of Clinical Nutrition*, 1995, 62, 1370S–1376S.

[69] Kohlmeier, L; Kark, JD; Gomez-Garcia, E; Martin, BC; Steck, SE; Kardinaal, AF; Ringstad, J; Thamm, M; Masaev, V; Riemersma, R; Martin-Moreno, JM; Huttunen, JK; Kok, FJ. Lycopene and myocardial infarction risk in the EURAMIC study. *American Journal of Epidemiology*, 1997, 146, 618-26.

[70] Kritenson, M; Ziedén, B; Kucinskienë, Z; Elinder, LS; Bergdahl, B; Elwing, B; Abaravicius, A; Razinkovienë, L; Calkauskas, H; Olsson, AG. Antioxidant state and mortality from coronary heart disease in Lithuanian and Swedish men: concomitant cross sectional study of men aged 50. *British Medical Journal*, 1997, 314, 629-633.

[71] Khachik, F; Carvallo, L; Bernstein, PS; Muir, GJ; Zhao, DY; Katz, NB. Chemistry, distribution and metabolism of tomato carotenoids and their impact on human health. *Experimental Biology and Medicine*, 2002, 227, 845-851.

[72] Kucuk, O; Sarkar, FH; Sakr, W; Djuric, Z; Pollak, MN; Khachik, F; Li, YW; Banerjee, M; Grignon, D; Bertram, JS; Crissman, JD; Pontes, EJ; Wood, DP Jr. Phase II randomized clinic trial of lycopene supplementation before radical prostatectomy. *Cancer Epidemiology, Biomarkers & Prevention,* 2001, 10, 861-868.

[73] Giovannucci, ERE; Liu, Y; Stampfer, MJ; Willet, WC. A prospective study of tomato products, lycopene, and prostate cancer risk. *Journal of the National Cancer Institute*, 2002, 94, 391-398.

[74] Rao, AV; Ray, MR; Rao, LG. Lycopene. *Advances in Food and Nutrition Research*, 2006, 51, 99-164.

[75] Mohanty, NK; Saxena, S; Singh, UP; Goyal, NK; Arora, RP. Lycopene as a chemopreventive agent in the treatment of high-grade prostate intraepithelial neoplasia. *Urologic Oncology*, 2005, 23, 383-385.
[76] van Breemen, RB; Xu, X; Viana, MA; Chen, L; Stacewicz-Sapuntzakis, M; Duncan, C; Bowen, PE; Sharifi, R. Liquid chromatography-mass spectrometry of cis- and all-trans lycopene in human serum and prostate tissue after dietary supplemenation with tomato sauce. *Journal of Agricultural and Food Chemistry*, 2002, 50, 2214-2219.
[77] Bowen, P; Chen, L; Stacewicz-Sapuntzakis, M; Duncan, C; Sharifi, R; Ghosh, L; Kim, HS; Christov-Tzelkov, K; van Breemen, R. Tomato sauce supplementation and prostate cancer: lycopene accumulation and modulation of biomarkers of carcinogenesis. *Experimental Biology and Medicine* (Maywood), 2002, 227, 886-893.
[78] Rao, LG. Tomato lycopene and bone health: Preventing osteoporosis. In: Rao AV, editor. *Tomatoes, lycopene and human health.* Scotland: Caledonian Science Press; 2006; 153-168.
[79] Silverton, S. Osteoclast radicals. *Journal of Cellular Biochemistry*, 1994, 56, 367–73.
[80] Liu, H-C; Cheng, R-M; Lin, F-H; Fang, H-W. Sintered beta-dicalcium phosphate particles induce intracellular reactive oxygen species in rat osteoblasts. *Biomedical Engineering: Applications, Basis and Communications*, 1999, 11, 259-264.
[81] Rao, LG; Krishnadev, N; Banasikowska, K; Rao, AV. Lycopene I-Effect on osteoclasts: Lycopene inhibits basal and parathyroid hormone-stimulated osteoclast formation and mineral resorption mediated by reactive oxygen species in rat bone marrow cultures. *Journal of Medicinal Food*, 2003, 6, 69-78.
[82] Kim, L; Rao, AV; Rao, LG. Lycopene II-Effect on osteoblasts: the caroteroid lycopene stimulates cell proliferation and alkaline phosphatase activity of SaOS-2 cells. *Journal of Medicinal Food*, 2003, 6, 79-86.
[83] Park, CK; Ishimi, Y; Ohmura, M; Yamaguchi, M; Ikegami, S. Vitamin A and carotenoids stimulate differentiation of mouse osteoblastic cells. *Journal of Nutritional Science and Vitaminology*, 1997, 43, 281-296.
[84] Paran, E. Reducing hypertension with tomato lycopene. In: Rao AV, editor. *Tomatoes, lycopene and human health.* Scotland: Caledonian Science Press; 2006; 169-182.
[85] Moriel, P; Sevanian, A; Ajzen, S; Zanella, MT; Plavnik, FL; Rubbo, H; Abdalla, DS. Nitric oxide, cholesterol oxides and endothelium-

dependent vasodialation in plasma of patients with essential hypertension. *Brazilian Journal of Medical and Biological Research*, 2002, 35, 1301-1309.

[86] Palan, P; Naz, R. Changes in various antioxidant levels in human seminal plasma related to immunofertility. *Archives of Andrology*, 1996, 36, 139-143.

[87] Foy, CJ; Passmore, AP; Vahidassr, MD; Young, IS; Lawson, JT. Plasma chain-breaking antioxidants in Alzheimer's disease, vascular dementia and Parkinson's disease. *Quarterly Journal of Medicine*, 1999, 92, 39-45.

[88] Schmidt, R; Fazekas, F; Hayn, M; Schmidt, H; Kapeller, P; Roob, G; Offenbacher, H; Schumacher, M; Eber, B; Weinrauch, V; Kostner, GM; Esterbauer, H. Risk factors for microangiopathyrelated cerebral damage in Austrian stroke prevention study. *Journal of Neurological Sciences*, 1997, 152, 15-21.

[89] Longnecker, MP; Kamel, F; Umbach, DM; Munsat, TL; Shefner, JM; Lansdell, LW; Sandler, DP. Dietary intake of calcium, magnesium and antioxidants in relation to risk of amyotrophic lateral sclerosis. *Neuroepidemiology*, 2000, 19, 210-216.

[90] Nahum, A; Hirsch, K; Danilenko, M; Watts, CK; Prall, OW; Levy, J; Sharoni, Y. Lycopene inhibition of cell cycle progression in breast and endometrial cancer cells in associated with reduction in cyclin D levels and retention of p27 (Kip1) in the cyclin E-cdk2 complexes. *Oncogene*, 2001, 20, 3428-3436.

[91] Matsushima, NR; Shidoji, Y; Nishiwaki, S; Yamada, T; Moriwaki, II; Muto, Y. Suppression by carotenoids of microcystin-induced morphological changes in mouse hepatocytes. *Lipids*, 1995, 30, 1029-1034.

[92] Sgherri, C; Pérez-López, U. Glutathione/ascorbate cycle and its interrelation with lipoic acid: could it explain why plants do not get cancer? In: Labrou N and Flemetakis E, editors. *Glutathione: Biochemistry, mechanisms of action and biotechnological implications*. New York: Nova Science Publishers, Inc; 2013; 233-262.

[93] Halliwell, B. Free radicals and other reactive species in disease. Encyclopedia of Life Sciences, 2001 Nature Publishing Group. www.els.net

[94] Calucci, L; Navari-Izzo, F; Pinzino, C; Sgherri, CLM. Fluidity changes in thylakoid membranes of durum wheat induced by oxidative stress: a

spin probe EPR study. *The Journal of Physical Chemistry B*, 2001, 105, 3127-3134.

[95] Sgherri, C; Pinzino, C; Navari-Izzo, F; Izzo, R; Quartacci, MF. Antioxidant potential in lipophilic and hydrophilic extracts from medicinal herbs (*Salvia officinalis* and *Echinacea angustifolia*). A comparison between assays based on electron paramagnetic resonance and spectrophotometry. *American Journal of Agricultural and Biological Sciences*, 2012, 7, 417-424.

[96] Lindberg Madsen, H; Bertelsen G. Spices as antioxidants. *Trends in Food Science and Technology*, 1995, 6, 271-277.

[97] Sgherri, C; Navari-Izzo, F; Pardossi, A; Soressi, GP; Izzo R. The influence of diluted seawater and ripening stage on the content of antioxidants in fruits of different tomato genotypes. *Journal of Agricultural and Food Chemistry*, 2007, 55, 2452-2458.

[98] Bohm, F; Edge, R; Land, EJ; McGarvey, DJ; Truscott, TG. Carotenoids enhance vitamin E antioxidant efficiency. *Journal of the American Chemical Society*, 1997, 119, 621-622.

[99] Calucci, L; Pinzino, C; Zandomeneghi, M; Capocchi, A; Ghiringhelli, S. Effects of γ-irradiation on the free radical and antioxidant contents in nine aromatic herbs and spices. *Journal of Agricultural and Food Chemistry*, 2003, 51, 927-934.

[100] Sgherri, C; Cecconami, S; Pinzino, C; Navari-Izzo, F; Izzo, R. Levels of antioxidants and nutraceuticals in basil grown in hydroponics and soil. *Food Chemistry*, 2010, 123, 416-422.

[101] Huang, D; Ou, B; Prior, RL. The chemistry behind antioxidant capacity assays. *Journal of Agricultural and Food Chemistry*, 2005, 53, 1841-1856.

[102] Prior, RL; Cao, GH. Analysis of botanicals and dietary supplements for antioxidant capacity: A review. *Journal of AOAC International*, 2000, 83, 950-956.

[103] Cao, GH; Prior, RL. Comparison of different analytical methods for assessing total antioxidant capacity of human serum. *Clinical Chemistry*, 1998, 44, 1309-1315.

[104] Re, R; Pellegrini, N; Proteggente, A; Pannala, A; Yang, M; Rice-Evans C. Antioxidant activity applying an improved ABTS radical cation decolorization assay. *Free Radical Biology and Medicine*, 1999, 26, 1231-1237.

[105] Lucarini, M; Pedrielli, P; Pedulli, GF; Valginigli, L; Gigmes, D; Tordo, P. Bond dissociation energies of the N-H bond and rate constants for the

reaction with alkyl, alkoxyl, and peroxyl radicals of phenothiazines and related compounds. *Journal of the American Chemical Society*, 1999, 121, 11546-11553.

[106] Ragaee, S; Abdel-Aal, ESM; Noaman, M. Antioxidant activity and nutrient composition of selected cereals for food use. *Food Chemistry*, 2006, 98, 32-38.

[107] Foti, MC; Daquino, C; Geraci, C. Electron-transfer reaction of cinnamic acids and their methyl esters with the DPPH radical in alcoholic solutions. *The Journal of Organic Chemistry*, 2004, 69, 2309-2314.

[108] Prior, RL; Wu, X; Schaich, K. Standardized methods for the determination of antioxidant capacity and phenolics in foods and dietary supplements. *Journal of Agricultural and Food Chemistry*, 2005, 53, 4290-4302.

[109] Espín, JC; Soler-Rivas, C; Wichers, HJ. Characterization of the total free radical scavenger capacity of vegetable oils and oil fractions using 2,2-Dyphenyl-1-picrylhydrazyl radical. *Journal of Agricultural and Food Chemistry*, 2000, 48, 648-656.

[110] Pérez-López, U; Pinzino, C; Quartacci MF; Ranieri, A; Sgherri, C. Phenolic composition and related antioxidant properties in differently-colored lettuce: a study by EPR kinetics. *Journal of Agricultural and Food Chemistry*, 2014, 62, 12001-12007..

[111] Wenli, Y; Yaping, Z; Zhen, X; Hui, J; Dapu W. The antioxidant properties of lycopene concentrate extracted from tomato paste. *Journal of American Oil Chemists' Society*, 2001, 78, 697-701.

[112] Everett, SA; Dennis, MF; Patel KB. Scavenging of nitrogen dioxide, thiyl, and sulfonyl free radicals by the nutritional antioxidant β-carotene. *The Journal of Biological Chemistry*, 1996, 271, 3988-3994.

[113] Mortensen, A; Skibsted, LH; Sampson, J; Rice-Evans, C; Everett, SA. Comparative mechanisms and rates of free radical scavenging by carotenoid antioxidants. *FEBS Letters*, 1997, 418, 91-97.

[114] Miller, NJ; Sampson, J; Candeias, LP; Bramley PM; Rice-Evans CA. Antioxidant activities of carotenes and xanthophylls. *FEBS Letters*, 1996, 384, 240-242.

[115] Staško, A; Brezová V; Biskupič, S; Mišík, V. The potential pitfalls of using 1,1-diphenyl-2-picrylhydrazyl to characterize antioxidants in mixed water solvents. *Free Radical Research*, 2007, 41, 379-390.

[116] Okazaki, S; Tomo, T; Mimuro, M. Direct measurement of singlet oxygen produced by four chlorin-ringed chlorophyll species in acetone solution. *Chemical Physics Letters*, 2010, 485, 202-206.

[117] Sgherri, CLM; Pinzino, C; Samaritani, E; Navari-Izzo, F. Activated oxygen generation from thylakoids: a novel spin trap. *Free Radical Research*, 1999, 31, S199-204.

[118] Endo, Y; Usuki, R; Kaneda, T. Antioxidants effects of chlorophyll and pheophytin on the autoxidation of oils in the dark. II. The mechanism of antioxidative action of chlorophyll. *Journal of the American Oil Chemists' Society*, 1985, 62, 1387-1390.

[119] Kraatz, M; Whitehead, TR; Cotta, MA; Berhow, MA; Rasmussen MA. Effects of chlorophyll-derived efflux pump inhibitor pheophorbide *a* and pyropheophorbide *a* on growth and macrolide antibiotic resistance of indicator and anaerobic swine manure bacteria. *International Journal of Antibiotics* 2014, 2014, 14 Article ID 185068,

[120] Hsu, CY; Chao PY; Hu SP; Yang CM. The antioxidant and free radical scavenging activities of chlorophylls and pheophytins. *Food and Nutrition Sciences*, 2013, 4, 1-8.

[121] Arimoto-Kobayashi, S; Inada, N; Nakano, H; Rai, H; Hayatsu, H. Iron-chlorophyllin-mediated conversion of 3-hydroxyamino-1-methyl-5H-pyrido[4,3-b]indole (Trp-P-2(NHOH)) into its nitroso derivatives. *Mutation Research*, 1998, 400, 259-269.

[122] Lai, J; Butler, MA; Matney, T. Antimutagenic activities of common vegetables and their chlorophyll content. *Mutation Research*, 1980, 77, 245-250.

[123] Sato, M; Fujimoto, I; Sakai, T; Aimoto, T; Kimura, R; Murata, T. Effect of sodium copper chlorophyllin on lipid peroxidation IX. On the antioxidative components in commercial preparations of sodium copper chlorophyllin. *Chemical and Pharmaceutical Bulletin*, 1986, 34, 2428-2434.

[124] Cahyana, H; Shuto, Y; Kinoshita Y. Antioxidative activity of porphyrin derivatives. *Bioscience, Biotechnology, and Biochemistry*, 57, 1993, 680-681.

[125] Hoshina, C; Tomita, K; Shioi, Y. Antioxidant activity of chlorophylls: Its structure-activity relationship. In: Garab G, editor. *Photosynthesis: Mechanisms and Effects*. Dordrecht, The Netherlands: Kluwer Academic Publishers; 1998; 4, 3281-3284.

[126] Tachino, N; Guo, D; Dashwood, WM; Yamane, S; Larsen, R; Dashwood, R. Mechanisms of the *in vitro* antimutagenic action of chlorophyllin against benzo[a]-pyrene: Studies of enzyme inhibition, molecular complex formation and degradation of the ultimate carcinogen. *Mutation Research*, 1994, 308, 191-203.

[127] Chernomorsky, S; Segelman, A; Poretz, RD. Effect of dietary chlorophyll derivatives on mutagenesis and tumor cell growth. *Teratogenesis, Carcinogenesis, and Mutagenesis*, 1999, 19, 313-322.

[128] Simonich, MT; Egner, PA; Roebuck, BD; Orner, GA; Jubert, C; Pereira, C; Groopman, JD; Kensler, TW; Dashwood, RH; Williams, DE; Bailey, GS. Natural chlorophyll inhibits aflatoxin B1-induced multi-organ carcinogenesis in the rat. *Carcinogenesis*, 2007, 28, 1294-1302.

[129] de Vogel, J; Jonker-Termont, DSML; van-Lieshout, EMM; Katan, MB; van der Meer, R. Green vegetables, red meat and colon cancer: chlorophyll prevents the cytotoxic and hyperproliferative effects of haem in rat colon. *Carcinogenesis*, 2005, 26, 387-393.

[130] de Vogel, J; Jonker-Termont, DS; Katan MB; van der Meer, R. Natural chlorophyll but not chlorophyllin prevents heme-induced cytotoxic and hyperproliferative effects in rat colon. *Journal of Nutrition*, 2005, 135, 1995-2000.

[131] Egner, PA; Munoz, A; Kensler, TW. Chemoprevention with chlorophyllin in indivuals exposed to dietary aflatoxin. *Mutation Research*, 2003, 523-524, 209-2016.

[132] Natsume, Y; Satsu, H; Kitamura, K; Okamoto, N; Shimizu, M. Assessment system for dioxin absorption in the small intestine and prevention of its absorption by food factors. *Biofactors*, 2004, 21, 375 - 377.

[133] Versantvoort, CHM; Oomen, AG; Van de Kamp, E; Rompelberg, CJM; Sips, AJAM. Applicability of an *in vitro* digestion model in assessing the bioaccessibility of mycotoxins from food. *Food and Chemical Toxicology*, 2005, 43, 31-40.

[134] Balder, HF; de Vogel, J; Jansen, MCJF; Weijenberg, MP; van den Brandt, PA; Westenbrink, S; van der Meer, R; Goldbohm, RA. Heme and chlorophyll intake and risk of colorectal cancer in the Netherlands Cohort Study. *Cancer Epidemiology, Biomarkers & Prevention*, 2006, 15, 717-725.

[135] Hsu, CY; Yang, CM; Chen, CM; Chao PY; Hu, SP. Effects of chlorophyll-related compounds on hydrogen peroxide induced DNA damage within human lymphocytes. *Journal of Agricultural and Food Chemistry*, 2005, 53, 2746-2750.

[136] Kamat, JP; Boloor KK; Devasagayam, TP. Chlorophyllin as an effective antioxidant against membrane damage *in vitro* and *ex vivo*. *Biochimica et Biophysica Acta*, 2000, 1487, 113-127.

[137] Sternberg, ED; Dolphin, D; Brückner, C. Porphyrin-based photosensitizers for use in photodynamic therapy. *Tetrahedron*, 1998, 54, 4151-4202.
[138] Nyman, ES; Hynninen, PH. Research advances in the use of tetrapyrrolic photosensitizers for photodynamic therapy. *Journal of Photochemistry and Photobiology B-Biology*, 2004, 73, 1-28.
[139] Singh, A; Singh, SP; Bamezai, R. Modulatory influence of chlorophyllin on the mouse skin papillomagenesis and xenobiotic detoxification system. *Carcinogenesis*, 1996, 17, 1459-1463.
[140] Singh, A; Singh, SP; Bamezai, R. Postnatal effect of arecoline on chlorophyllin-modulated hepatic biotransformation system enzymes in suckling neonate lactating mice. *Teratogenesis, Carcinogenesis and Mutagenesis*, 1996, 16, 89-96.
[141] Yun, CH; Jeong, HG; Jhoun, JW; Guengerich, FP. Non-specific inhibition of cytochrome P450 activities by chlorophyllin in human and rat liver microsomes. *Carcinogenesis*, 1995, 16, 1437-1440.
[142] Diaz, GD; Li, Q; Dashwood, RH. Caspase-8 and apoptosis-inducing factor mediate a cytochrome c–independent pathway of apoptosis in human colon cancer cells induced by the dietary phytochemical chlorophyllin. *Cancer Research*, 2003, 63, 1254-1261.
[143] Chan, JY; Tang, PM; Hon, PM; Au, SW; Tsui, SK; Waye, MM; Kong, SK; Mak, TC; Fung, KP. Pheophorbide a, a major antitumor component purified from *Scutellaria barbata*, induces apoptosis in human hepatocellular carcinoma cells. *Planta Medica*, 2006, 72, 28-33.
[144] Chiu, LC; Kong, CK; Ooi, VE. The chlorophyllin-induced cell cycle arrest and apoptosis in human breast cancer MCF-7 cells is associated with ERK deactivation and cyclin D1 depletion. *International Journal of Molecular Medicine*, 2005, 16, 735-740.
[145] Demmig-Adams, B; Adams WW. Food and photosynthesis: antioxidants in photosynthesis and human nutrition. *Science*, 2002, 298, 2149-2153.

Reviewed by

Prof. Annamaria Ranieri, Dipartimento di Scienze Agrarie, Alimentari e Agro-ambientali, University of Pisa, Via del Borghetto 80, I-56124 Pisa, Italy.

In: Lycopene
Editor: Jacob R. Bailey
ISBN: 978-1-63117-927-3

Chapter 3

LYCOPENE EXTRACTION AND ANALYSIS

Dandara Fidelis Escoto[1], Bruna Piaia Ramborger[1], Mateus Cristofari Gayer[1], Daniela Teixeira Rodrigues[1], Elton Luis Gasparoto Denardin[2], Rafael Roehrs[1] and Miguel Roehrs[3,*]

[1]Federal University of Pampa, Interdisciplinary Research Group on Teaching Practice, Natural Sciences, Bagé, Brazil
[2]Federal University of Pampa, Physical and Chemical Research and Natural Products Laboratory, Post Graduate Program in Biochemistry, Bagé, Brazil
[3]Federal University of Santa Maria, University Hospital of Santa Maria, Post Graduate Program in Pharmacology, Integrated Center for Development of Laboratory Analysis, Santa Maria, Brazil

ABSTRACT

Carotenoids are substances produced by fruits and vegetables, which are responsible for their red and orange tones. Lycopene belongs to the group of carotenoids and currently appears as one of the most potent antioxidants. Most carotenoids present linear structure with 40 carbons with 3 to 15 conjugated double bonds, which impart their property of absorbing light at wavelengths between 400 and 500 nm. Lycopene is a carotenoid without pro-vitamin A activity, lipid-soluble, composed of eleven conjugated bonds and two double bonds unconjugated. Lycopene is considered the carotenoid that has the highest scavenging capacity of

* email address: miguelroehrs@yahoo.com.br

singlet oxygen, possibly due to the presence of two double bonds unconjugated, which offers greater reactivity. For these reasons it is important to determine the amount of lycopene on different samples. Lycopene is the predominant carotenoid in human plasma and tissues, being found in a limited number of red colored foods as tomatoes and their products, guava and watermelon. The first step is the extraction of lycopene from the samples. This procedure can be made with different types of solvent and different kinds of extraction processes like liquid-liquid extraction (LLE), Solid-liquid extraction, dispersive solid phase extraction(d-SPE), solid phase extraction (SPE). The amount of lycopene after the extraction has been determined in food or biological samples by many analytical methods, such as UV -Visible spectrophotometry, liquid chromatography connected to electrospray-ionization (LC/ESI-MS), reverse phase liquid chromatography (RP-LC), supercritical fluid chromatography (SFC), matrix assisted desorption ionization (MALDI) and especially LC with spectrophotometric detection.

Keywords: cruise tourism, Scotland, limitations, tourism development

1. Introduction

Carotenoids are substances produced by fruits and vegetables that give red and orange tones. At moment were described more than 600 of these compounds in nature [1]. The majority of the carotenoids is linear structure with 40 carbons with 3 to 15 conjugated double bonds, which impart the property of absorbing light at wavelengths between 400 and 500 nm. Carotenoids containing only carbon and hydrogen in their structures are called carotenes, for example, β-carotene, α-carotene and lycopene.

Lycopene is present in plasma and tissues with large variations in their distribution. The presence of carotenoids in human tissue has been reported since 1990; it is known that these carotenoids and their metabolites are present in serum or accumulate in tissues such as liver, lung, breast, cervical and skin. Among the carotenoids, lycopene is one of the most abundant in the human body, with its high concentration mainly due to the consumption of food sources. The human body is unable to synthesize carotenoids, so they must give obtained solely by diet. Lycopene can be found in a limited number of foods; the tomato and its derivatives are the best dietary contributions, but are good sources of this element also papaya, red guava, cherry and watermelon [2,3,4]. The problem that affects the tomato crop is the short shelf life in post-harvest, because of their susceptibility to different types of pests [5]. Due to this factor, is used in large-scale pesticides and fertilizers, disregarding nuisance and waste to own tomatoes, which ends up suffering an imbalance of nutrients, affecting the amount of lycopene. Similarly to this problem affects tomato, also affects other crops. Thus it is necessary to determine the amount of lycopene in foods raw or processed, seeking food sources that contemplate the needs of the daily diet [6].

Lycopene currently appears as one of the most potent antioxidants been suggested in the prevention of carcinogenesis and atherogenesis for protecting molecules such as lipids, low density lipoproteins (LDL), proteins and DNA. Epidemiological studies before 1995 revealed that the intake of tomatoes and blood lycopene level are inversely associated with the risk of developing cancers at several anatomical sites, including the prostate gland, stomach, and lung. May be that it may provide protection against cancer and other degenerative diseases, influenced by free radical reactions[7,8]. All-trans-lycopene is an unsaturated acyclic hydrocarbon with chemical formula C40H56 and molecular weight of 536.85. Its Chemical Abstract Service (CAS) Number is 502-65-8. The chemical name of all-trans-lycopene is (all-E)-2,6,10,14,19,23,27,31-octamethyl-2, 6, 8, 10, 12, 14, 16, 18, 20, 22, 24,26, 30-dotriacontatridecaene. Common names include Ψ,Ψ-carotene, all- trans-lycopene, and (all-E)-lycopene. The structural formula of all-trans-lycopene is shown in Figure 1.

In food, the tomatoes and its products are considered as one of the best sources of lycopene. The total lycopene content in tomatoes varies between 90 and 190 μg/g fresh weight [9]. Lycopene is found predominantly in the chromoplast of plant tissues. In tomatoes, lycopene biosynthesis increases dramatically during the ripening process, as chloroplast undergoes transformation to chromoplast. Globulous chromoplast containing mainly β-carotene is found in the jelly part of the pericarp while chromoplast in the other part of the pericarp contains voluminous sheets of lycopene [10].

2. Extraction Techniques and Analyses Methods for Lycopene in Vegetables Extracts

Different extraction and quantitation methods for lycopene are recorded in the literatures since 1940. Determination of total carotenoids can be made with spectrophotometric analysis: total pigments were then quantified by absorption spectrum recorded at a wavelength of 450 nm [9], and the mathematical expression described by Gross [10], whereas the average absorption coefficient for 2500 total carotenoids. An experimental plan for extracting carotenoids, and lycopene may be described as follows: 200 g samples were taken for each product (tomato, guava and acerola) which was then subjected to the multiprocessor to be ground to homogeneity with 50 ml of water to form a paste. Then 40 g were taken from each sample of fruit, which were submitted to exhaustive extraction of carotenoids from the pulp by successive washings with acetone. Immediately after the addition of the ether extract was taken in a separatory funnel where the lower phase (acetone) was discarded. To determination the concentration of lycopene in the fruits was also obtained by spectrophotometric analysis. The reading on the spectrophotometer lycopene content was made at a wavelength of 470 nm [9]. For lycopene in petroleum ether value of the extinction coefficient is 3450 [9]. Calculated using the formula: mg / g = (V x A x 1.000.000) ÷ (M x A1 x 100) where A = absorbance of the solution at the wavelength, V = final volume of the solution, A1 is the extinction coefficient or molar absorption coefficient of a pigment in a certain specific solvent, and M = mass of the sample taken for analysis.

A alternative method for lycopene extraction are described by Silva (2001) use acetone and petroleum ether. The extraction with acetone was carried out in three stages, starting from the same amount of sample [11]. Aliquots of sample of the concentrated slurry were ground with solvent using four parts of acetone for one part sample and left in soaking for 24 hours in sealed glass containers under ambient light and temperature (22 ± 2 °C). After steeping, the extracts are filtered vacuum and waste reextracted with the same sample rate: solvent in two more steps, with four replications. The filtrates obtained from each extract were separated in a separatory funnel with petroleum ether and washed with distilled water four times. Subsequently, the ether extracts were transferred to a round bottom flasks and concentrated under reduced pressure at a temperature of 30-35°C [11].

For total carotenoids in sample or lycopene, FAO [12] indicate a method using spectrophotometer and liquid chromatography. To sample preparation, a introduce a representative sample of the tomato extract into a vial and close it. Place the vial in a water bath at 50 °C for 30 minutes. Stir the solution using a glass rod. Load accurately the sample into each of three volumetric flasks and add BHT solution. Solution should be protected from light and added dichloromethane to each flask. Homogenize the solutions using an ultrasonic bath, cool the solutions to room temperature and bring each to volume with dichloromethane and mix. Transfer the last solution to separate amber volumetric flasks. Bring each to volume with the diluent solution and mix well.

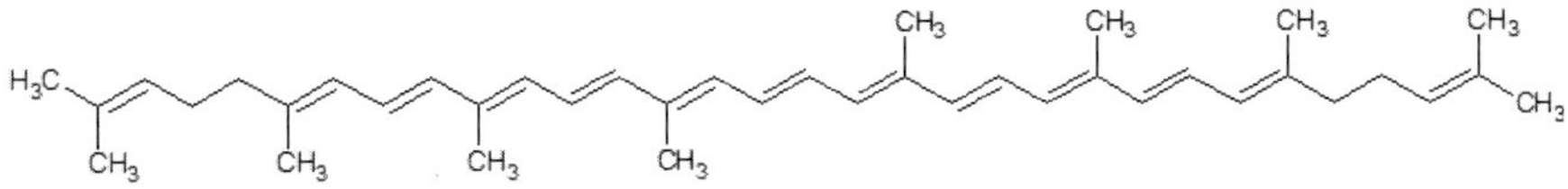

Figure 1. All-trans-lycopene.

Another way to extract can be the extraction method by GROSS, this standard method for the extraction of carotenoids. Before removal can be added a antioxidant BHT to avoid oxidation or enzymatic degradation. In tissues containing a high percentage of water is facilitated extraction with polar organic solvents such as acetone, methanol or ethanol. In dry samples can be made to the extraction using ethyl ether or petroleum ether, with the aid of a grinder in order to facilitate extraction, and the procedure repeated until the residue become colorless [13]. To analyze the total carotenoids amount by espectrophometric assay, a volumetric pipette are used to transfer 2 ml of solution previous prepared to an amber volumetric flask. Add ethanol, bring to volume with petroleum ether and mix well. Using a suitable UV/VIS spectrophotometer and 1-cm sample cells with covers, scan the spectrum of Solution G from 550 to 300 nm, using petroleum ether as the reference blank and measure the absorbance at the absorbance maximum (approximately 472 nm).

In the chromatographic analysis the parameters are a HPLC system with a UV/VIS detector or a diode array detector and auto sampler, Detector at 472 nm, Column: RP-C8 (250 x 4.6 mm, 5 μm) with mobile phase: acetonitrile: methanol: dichloromethane :n-hexane: N-ethyl-diisopropylamine 850:100:25: 25: 0.5 (v/v/v/v/v). Mix well and sonicate in an ultrasonic bath, flow rate: 0.7 ml/min, injection volume: 10 μl with a run time: 12 min [12]. Silva [11] presents the lycopene extraction using three solvent mixtures and subsequent analysis by spectrophotometry according Gross in 1991. In these techniques are used as solvents acetone, hexane and ethanol, in the following proportions: hexane-acetone (7:3); hexane-ethanol (1: 1) and acetone-hexane (1: 1). The extraction are performed in three steps (reextractions) taking the same amount of sample. Aliquots of concentrated sample or partially dehydrated tomato pulp were triturated with solvents, using 4 parts of solvent for each part of the sample left in soaking for 24 hours in sealed glass container protected from light, at room temperature (22 ± 2 °C). After steeping, the extracts were filtered vacuum and waste reextracted with the same sample rate: solvent in two more steps, with four replications. The filtrate of each extract was separated by separatory funnel with petroleum ether and washed with distilled water four times, with the exception of extract, which was washed with water directly without the use of petroleum ether. Thereafter, ether and hexane extracts were transferred into round bottom flasks and concentrated under reduced pressure at 30-35 ºC temperature. Then, each flask was washed with ethyl alcohol and 10 ml of hexane for later reading absorbance at 472 nm, the reading obtained at a wavelength between 400 and 520 nm in the ultraviolet/visible spectrophotometer.

Tomato pulp, the efficiency of extraction solutions in descending order was extracted acetone, hexane-ethanol (1: 1) and acetone-hexane (7: 3). The hexane solutions, acetone-hexane (1: 1) and ethanol were the ones that had less power extractor, the latter being the worst performance. For partially dehydrated tomato, the solvent hexane-ethanol (1: 1), acetone and acetone-hexane (1: 1) showed greater power extraction, followed by acetone-hexane (7: 3) and hexane. The ethanol solvent led to lower extraction performance.

A different method to quantify lycopnene using chromatohgraphy are described by Thadikamala et al. [14]. They used the Supelcosil LC-18 analytical column 3 μm, 150 × 4.6 mm i.d (Supelco, USA), the mobile phase consisted of acetonitrile, methanol and tetrahydrofuran (70:25:5, v/v) and was delivered at flow rate of 1.0 mL/min. The retention time for lycopene is 14.2 min. The wavelength are set at 475 nm. The chromatographic system is a Shimadzu LC-10 AS system (controller, pumps, injector equipped with a 20 μL sample loop) and an SPD-10A UV-Visible detector.

Seo et al. [15] show different ways to extract lycopene from pumpkin, they use liquid-liquid extraction (LLE) and supercritical fluid extraction. In LLE, homogeneous freeze-dried pumpkin samples are mixed with ethanol (95%) containing BHT to denature proteins, vortexed for 60 seconds, then mixed with an equal volume of hexane. The hexane layer is remove and save in 12 mm X 75 mm borosilicate test tubes, and the extraction repeated three times. Each hexane extract was dried under a stream of nitrogen. The extracts are resuspended in solvent B. Samples and standards are stable under these conditions for up to one week. The HPLC method use was mobile phase solvents were made by mixing acetonitrile, tetrahydrofuran and methanol, then adding 1% ammonium sulfate in HPLC grade water and 0.0005% β-hydroxytoluene as an antioxidant. Solvent A contained acetonitrile–tetrahydrofuran–methanol–1% aqueous ammonium sulfate solution (85:5:5:5, v/v/v/v), and solvent B contained acetonitrile–tetrahydrofuran–methanol–1% aqueous ammonium sulfate solution (55:35:5:5, v/v/v/v). Each solvent was prepared fresh before use. HPLC standards were prepared by dissolving approximately 1 mg standard into 10 mL toluene–ethanol (1:99, v/v).

Standards were prepared by mass, and stored in the dark at 4 ºC for up to two weeks before use.

Pumpkin carotenoids analysis are measured by reversed-phase liquid chromatography using diode array detection. Chromatography was run as a gradient at 1.0 mL/min with a Prodigy 5μ C_{18}ODS 3 100 A pore 250 mm X 4.6 mm reversed-phase column (Phenomenex, Torrance, CA, USA). Total run time was 48 min. The gradient changed from 5% to 95% solvent B, as follows: 0–10.0 min, 5% solvent B; 10.0–29.0 min increasing linearly to 95% solvent B; 29.0–35.9 min, maintaining 95% solvent B; 35.9–36.0 min, abruptly decreasing to 60% solvent B; maintaining 60% solvent B from 36.0 to 44.9 min; then abruptly decreasing to 5% solvent B at 45 min. The column re-equilibrated with 5% solvent B from 45 to 48 min. Serum carotenoid concentrations were measured by diode array detection set at 452 nm. Supercritical fluid extraction is less toxic and less labor intensive than liquid–liquid extractions. Extraction recovery depends on the vegetable matrix, but has sometimes been excellent especially for β-carotene. Seo et al., [15] method did not produce complete recovery, but recovery was similar or better than milder liquid–liquid extraction methods. SPE is a simple, cost-effective and versatile method. A number of SPE methods employing different sorbents such as C18, diol, silica gel, and alumina for carotenoids analysis and purification have been reported [16-18].

For carotenoids like lycopene Shen et al. [19] use an C18 and C30 cartridges were conditioned with 3 mL of dewatering acetone and 3 mL of a mixed solvent (isopropanol + ethyl acetate+water,1+1+1,v/v/v) in sequence, respectively. The sample solution was then loaded onto the cartridge at the flow rate of 1 mL/min. The cartridge was washed by 5 mL of a mixed solvent (ethanol + water, 1 + 1, v/v), and eluted by 1 mL of acetone in sequence. The eluate are collected, and dried under a gentle nitrogen stream. The residue of sample solution is redissolved in ethanol, mixed solvent (ethanol + ethyl acetate, 1 + 1, v/v) each containing 0.05 μg/mL I.S., respectively. The final solution was ready for the HPLC analysis. The separation of analytes was carried out on a Prontosil C30 column (5μm, 250 mm X 4.6 mm, Prontosil, Bischoff Chromatography, Germany). Flow rate was set at 1.0 mL/min and the column temperature was kept at 15 ºC. Injection volume was 30 μL. The high retentions of β-carotene are achieved with the C18 and C30 cartridges when using the mixed solvent of isopropanol + ethyl acetate + water (1+1+1, v/v/v) as the loading solvent. Comparing with a traditional LLE method, the optimized SPE method is more efficient for the extraction of β -carotene in sample, time-saving, and environmentally. Lycopene can be extracted from plants and zymotic liquid via conventional solvent extraction, super critical fluid (CO_2) extraction (SCFE) and supersonic assisted extraction (SAE).

Conventional method requires long processing time with low efficiency while SCFE represents non-organic solvents residing as advantage but the equipment is expensive and the energy consumption is very high. Now is know that ultrasound and microwave radiation could accelerate the extracting process and this may improve bioactive compound extraction. Microwave assisted extraction heat the extracts quickly and accelerate the extraction process for adsorption and desorption of the targeted compounds from matrix, while its disadvantage is in homogeneous heating. Tomato puree are centrifuged at 3000 rpm by horizontal scroll-bowl centrifuge for 5 min. The supernatant was removed and the sedimentary is collected to next step. After centrifugation, the moisture content of the dehydration treatments are adjusted by vacuum-drying at 60 °C and 0.01 MPa pressure for 8–10 h. The final moisture content of the samples in all experiments was controlled at 78 ± 0.2% (W/W). The sample was stored in the refrigerator at 2–8°C until use. Lianfu and Zelong show others extractions forms like ultrasonic and microwave assisted extraction (UMAE) and ultrasound assisted extraction (UAE). UMAE experiment was carried out with ultrasonic and microwave extracting apparatus. An open microwave with maximal power of 800 W at a frequency of 2450 MHz, and an ultrasonic transducer with a fixed power of 50 W at a frequency of 40 KHz is used simultaneously to extract lycopene from tomato paste: transfer the sample into the flask, add proper volume of ethyl acetate and then transfer the flask into the chamber of the apparatus and connect it with condensing tubes. Finally, close the door of chamber and set the program of the parameters (microwave power and extraction time) according to the experimental design. When the extraction complete, the flask is removed from apparatus and the mixture is filtrated, then the volume of the filtrate and lycopene content is measured. The Ultrasonic assisted extraction are carried out in an open rectangular ultrasonic processor with maximal power of 300 W at 40 KHz frequency while operating at 50 W in practical process: take tomato paste into flask, add proper volume of ethyl acetate, agitate the mixture, connect the flask with fitted condensing works and then immerse the flask in the ultrasonic water bath, the temperature of the water bath is controlled automatically. When the extraction finish, the flask was taken out and cooled to room temperature, the mixture is filter, and finally the volume of the filtrate and lycopene content is measured. To avoid lycopene decomposing during the extraction, these processes should be done in darkroom. To quantify the amount of lycopene, Lianfu and Zelong [20], determined according to the absorbance measured at 502 nm via UV–VIS Spectrophotometer. The authors conclude that the comparison of these two methods showed that UMAE overcomes the short comings of UAE and will be a more attractive extract method in the future. Liquid chromatography (LC) coupled with UV-vis detection seems to be the preferred method for analysis of ascorbic acid, lycopene, and β-carotene. The overwhelming majority of these methods involve the photodiode array detector, which allows examining all wavelengths at the chromatographic peak to confirm the presence of a previously identified compound by retention time.

However, food samples require the application of more reliable and selective methods for definitive peak identification that only can be achieved by mass spectrometry (MS) [21]. Briefly, it consists of mixing 1 g of the vegetal in a polytron mixer with $MgCO_3$ and a mixture of tetrahydrofuran/methanol (1:1) (v/v). The solvent is removed by filtration and the filter cake was re-extracted again. Then, aqueous NaCl solution (10%) are add and a liquid-liquid extraction is carried out with 50 mL of petroleum ether (twice). The organic phases are joined, and the solvent is removed under vacuum in a rotatory evaporator until almost dry and then just to the point of dryness with a slight N_2 stream. The residue is redissolved in 10 mL of initial mobile phase. This solution is use for quantification by the standard additions method, as in the previous determination, and injected into LC-MS. A liquid chromatography system is equipped with autosampler, degasser, and heater column. The mass spectrometer system is a ZQ 2000 single quadrupole. The compounds are separated by two coupled columns, first a Symmetry C18 75x4.6 mm i.d. 3.5μm and then an Atlantis dC18 150x2.0 mm i.d. 5 μm, both from Water. A mobile phase composition of 70% methanol (0.005% acetic acid) and 30% acetic acid 0.05% in isocratic mode was used for ascorbic acid determination, while for carotenoid analysis a mixture of methanol, tetrahydrofuran, and acetonitrile (60:30:10 v/v/v) was used. Both columns are set to 30°C during all runs, and the injection volume is 100 μL for carotenoids.

Chemical ionization atmospheric pressure (APCI) in positive mode is used for the target carotenoids. The temperature source set at 130 °C, the flows for desolvation and cone gas are 400 L/h and 60 L/h, respectively, and the capillary set to 3.5 kV. Selective ion monitoring (SIM) was acquired by previous infusion of each standard in full scan mode at a concentration of 15 mg/L. A green alternative analysis to quantified the amount of lycopene in food are Fourier transform infrared (FT-IR) in combination with multivariate analysis, that offers a powerful and rapid technique for the analysis of agricultural and food products. Spectral bands arising from functional group vibrations of organic molecules in the mid-infrared (4000-700 cm^{-1}) region may be associated with specific chemical entities with known assignment in most cases. In addition, the mid-IR spectra allows for the chemically based discrimination of organic constituents, producing distinct and reproducible biochemical fingerprints that reflect the total biochemical composition of the sample, with bands due to major constituents such as lipids, proteins, nucleic acids, polysaccharides, and phosphate-carrying compounds. Genetic resources that include naturally occurring variants of genes affecting both the structure and regulation of key enzymes in the carotenoid biosynthesis pathway make tomatoes an excellent model food for nutritional studies.

The development of improved analytical techniques that allow the efficient measurement of diverse carotenoids in very large genetically characterized populations is greatly needed. Attenuated total reflection Fourier transform infrared spectroscopy (ATR-IR) is one of the most powerful and widely used sampling tools for collecting IR spectra of biological materials. The most important feature of the ATR accessory is the evanescent field, which develops during the reflection of radiation at the interface of a material with a high refractive index (ATR crystal) and a material with a low refractive index (sample). Attenuation of this electric field by functional groups in the lower refractive index material results in a spectrum analogous to an absorbance spectrum [22]. The ATR-IR analysis are fast and low cost, because in unnecessary sample preparation. An example of lycopene analysis by infrared are described in de Nardo et al., [22]. They use tomato juice aliquots (5 mL) mixed with methanol (5 mL) and centrifuged for 5 min at a speed of 3000 rpm. The supernatant was discarded, and the pellet (juice solids) was placed onto an AP15 glass fiber filter disk; samples were pressed onto the ATR-IR diamond crystal using a high-pressure clamp for infrared spectra acquisition. Each tomato juice sample was analyzed in duplicate (two cans), and four independent measurements of the paste were collected per sample, giving a total of eight spectra per juice sample.

If is necessary is possible clear the sample, using a sample preparation technique like that used too by de Nardo et al., [22] to compare direct analysis and analysis with cleanup. Extraction of lycopene and β-carotene from tomato juices was done as reported by Nguyen and Schwartz [23] with some modifications. A controlled light environment was used to prevent isomerization and photo-degradation. Tomato juice aliquots (5 mL) were mixed with methanol (5 mL) and centrifuged for 5 min at a speed of 3000 rpm. The supernatant was discarded, and 0.2 g of the pellet was re-suspended with a mixture of a 1:1:2 acetone/ethanol/hexane solution (20 mL volume) and continuously shaken for 15 min at 180 rpm. Deionized distilled water (3 mL) was later added and shaken continuously for 5 min at 180 rpm to induce phase separation. The hexane (top) layer was collected and brought up to 10 mL in a volumetric flask. An aliquot (3 mL each) was dried under nitrogen, re-suspended in chloroform (60μL), and applied onto the ATRIR Zn/Se crystal for infrared spectral acquisition. The remaining aliquots (2 X 3 mL) were dried under nitrogen and immediately analyzed by HPLC. Two cans for each tomato variety sample were opened, and duplicate or triplicate carotenoid extractions were performed with two independent measurements made by ATR-IR spectroscopy, giving a total of 8-12 spectra per juice sample.

In 2014 Stinco and colleagues [24] proposed a separation and quantification of different carotenoids, including lycopene including using rapid resolution liquid chromatography method. To this 10 mg of homogenized freeze-dried powder from the samples were used for the extractions. Due to the then importance of sampling in the analytical procedures and the small amount of dried materials used for this micro-extraction proto-col, it was important to start from especially representative large quantities of the materials. Appropriately taken to ensure representativeness and homogenize it thoroughly upon freeze-drying to keep the representativeness. The powder was gently mixed with 1 ml of the Milli Q-water and vortexed and then centrifuged to remove the aqueous phase at 18.000 X g for 3 min. Subsequently, 1 ml of extracting solvent (hexane / acetone, 1: 1 v / v) was added, the mixture was vortexed and centrifuged Then for 3 min at 18,000 X g. After recovering the colored fraction, the further 500 μL of extracting solvent was added, and the mixture was vortexed and finally spun. These operations were repeated until color exhaustion. The pooled organic colored fractions were evaporated to dryness in a vacuum concentrator at a temperature below 30ºC and stored under N_2 at -20 ºC until analysis. The dry residue was re-dissolved in acetonitrile prior to their injection in the RRLC system. The RRLC analyzes were carried out on a liquid chromatography system equipped with a diode-array detector, which was set to scan from 200 to 770 nm. The C18 Poroshell 120 column (2.7 uM, 5 cm × 4.6 mm) kept at 28◦C was used as stationary phase. The injection volume was set at 10-20 μL. The mobile phase was pumped at 1 mL / min and consisted of three solvents: solvent A, acetonitrile, solvent B, and solvent C methanol, ethyl acetate. The gradient elution was linear 0 min 85% A + 15% B; 5 min, 60% A + 20% B + 20% C; 7 min 60% A + 20% B + 20% C; 9 min 85% A + 15% B; 12 min 85% A + 15% B.

REFERENCES

[1] Almeida-Muradian, LB; Penteado, MVC. Carotenóides. In: Vitaminas: aspectos nutricionais, bioquímicos, clínicos e analíticos. *Manole*, 2003, 1-52.

[2] Khachik, F; Carvalho, L; Bernstein, PS; Muir, GJ; Zhao, DY; Katz, NB. Chemistry, distribuition, and metabolism of tomato carotenoids and their impact on human health. *Exp. Biol. Med.*, 2002, v.227, 845-851.

[3] Bramley, PM. Is lycopene beneficial to human helth? *Phytochemistry*, 2000, v.54, n. 3, 233-236.

[4] Shami, NJI; Moreira, EAM. Licopeno como agente antioxidante. *Ver. Nutr.*, 2004, v.17, n. 2, 227-236.

[5] Moreira, GC; Vieites. RL; Campos, AJ; Júnior, ED. Efeito da radiação gama na sanitização do tomate minimamente processado à vácuo. *Rev. Hig. Alim*, 2005, v.19, 132.

[6] Ferreira. SMR. Características de qualidade de tomate de mesa (Lycopersicon esculentum Mill.) cultivado nos sistemas convencional e orgânico comercializado na região metropolitana de Curitiba. p. 249, Curitiba, 2004.

[7] Levy, J; Bosin, E; Feldman, B. Lycopene is a more potent inhibitor of human cancer cell proliferation than either alpha-carotene or beta-carotene. *Nutrition and Cancer.*, 1995, v. 24, 257–266.

[8] Stahl, W; Sies. H. Lycopene: a biologically important carotenoids for human. *Archives of Biochemistry and Biophysics*, 1996, v. 336, 1-9.

[9] Rodriguez-Amaya, D. A Guide to Carotenoids Analysis in Food. Washington: International Life Sciences Institute Press, 2001, p.64.

[10] Gross, J. *Pigments in fruits.* London, UK : Academic Press. 1987.

[11] Silva, AG. Extração e estabilidade dos carotenoids obtidos de tomate processado (Lycopersyon esculentum Mill). Dissertação de mestrado. Universidade de Viçosa, MG, 2001.

[12] Food and Agriculture Organization (FAO). Carotenoids. <http:/ /faostat.fao.org > acessed in 12/10/2015.

[13] Gross, J. Pigments in vegetables: chlorophylls and carotenoids. New York: Van Nostrand Reinhold, 1991, p. 351.

[14] Sathish, T; et al. HPLC Method for the Determination of Lycopene in Crude Oleoresin Extracts. *Asian Journal of Chemistry*, 2009, v. 21, n. 1, 139-148.

[15] Seo, JK; et al. Extraction and chromatography of carotenoids from pumpkin. *Journal of Chromatography A*, 2005, v.1073, 371–375.

[16] Gatti, R; Gioia, MG; Cavrini, V. Analysis and stability study of retinoids in pharmaceuticals by LC with fluorescence detection. *J. Pharm. Biomed*, 2000, v.1, 147-159.

[17] Mateos, R; García-Mesa, JA. Direct spectrophotometric determination of bitterness in virgin olive oil without prior isolation by pH gradiente. *Anal. Bioanal. Chem*, 2006, v.30, 9615-9619.

[18] Britton, G; Charlwood, BV; Banthorpe, DV; Dey, PM; Harbone, JB. Methods in Plant Biochemistry, Academic Press, London, San Diego, CA, 1991, p. 473.

[19] Shen, Y; etal. Solid-phase extraction of carotenoids. *J. Chromatogr. A*, 2009, v.1216, 5763–5768.

[20] Lianfu, Z; Zelong, L. Optimization and comparison of ultrasound/microwave assisted extraction (UMAE) and ultrasonic assisted extraction (UAE) of lycopene from tomatoes. *Ultrasonics Sonochemistry*, 2008, v.15, 731–737.

[21] Frenich, AG; et al. Determination of ascorbic acid and carotenoids in food commodities by liquid chromatography with mass spectrometry detection. *J. Agric. Food Chem*, 2005, v. 53, 7371−7376.

[22] De Nardo, T; et al. Rapid and Simultaneous determination of Lycopene and β-Carotene Contents in Tomato Juice by Infrared Spectroscopy. *J. Agric. Food Chem*, 2009, v.57, 1105–1112.

[23] Nguyen, ML; Schwartz, SJ. Lycopene stability during food processing. *Proc. Soc. Exp. Biol*, 1998, v.218, 101–105.

[24] Stinco, CM; et al. Development and validation of a rapid resolution liquid chromatography method for the screening of dietary plant isoprenoids: Carotenoids, tocopherols and chlorophylls. *Journal of Chromatography A*, 2014, v.1370, 162–170.

In: Lycopene
Editor: Jacob R. Bailey
ISBN: 978-1-63117-927-3

Chapter 4

CHEMISTRY, SOURCES AND BENEFITS OF LYCOPENE

Cecília Elenir Rocha[1], Rafael Roehrs[1] and Miguel Roehrs[2]
[1]Federal University of Pampa, Interdisciplinary Research Group on Teaching Practice, Natural Sciences
[2]Federal University of Santa Maria, University Hospital of Santa Maria, Post Graduate Program in Pharmacology, Integrated Center for Development of Laboratory Analysis

ABSTRACT

The lycopene ((6E, 8E, 10E, 12E, 14E, 16E, 18E, 20E, 22E, 24E, 26E)-2, 6, 10, 14, 19, 23, 27, 31-octamethyldotriaconta-2, 6, 8, 10, 12,14, 16, 18, 20, 22, 24, 26, 30-tridecaene) is part of the over 600 carotenoids group which exists in nature, the class of carotenes, due to its structure that is formed only of carbon and hydrogen. It is also one of the 25 carotenoids which can be found in human plasma and tissues. Its chemical formula is $C_{40}H_{56}$, and its atoms are arranged in a long chain acyclic, with 13 double bonds, and 11 of these bonds being conjugated bonds. Due to the absence of the β-ionone ring at the end of its molecule structure, it has no pro-vitamin A activity as other carotenes; it is lipo-soluble or insoluble in water, methanol and ethanol. It is a plant pigment with reddish color, being regarded as the carotenoid with the largest capacity of scavenging singlet oxygen and the ability to reduce the rates of free radicals, even being able to make them inactive. As a result of these aspects, it performs an important antioxidant action in the body and is associated therewith to research involving the prevention and treatment of different pathologies. Studies show that the activity of this carotenoid may reduce the occurrence of various types of cancer, such as esophageal, gastric, prostate and lung, and assist in the treatment of pancreatic cancer, colon and rectum, and it is inversely associated with the risk of myocardial infarction. Lycopene can be extracted from some fruits and vegetables, the main sources are tomato (*Lycopersicon esculentum*), red guava (*Psidium guajava*), watermelon (*Citrullus lanatus*), papaya (*Carica papaya*) and surinam cherry (*Eugenia uniflora*). The tomato (in its natural form) is the one with the highest concentration of lycopene in accordance with the degree of ripening, available as trans isomers (*trans*-lycopene) to be more stable in nature. However, the *cis* isomers of lycopene are most commonly found in plasma and absorbed by the human body, because they have greater solubility in micelles. This *cis-trans* isometry conversion is facilitated by thermal processing because it facilitates the rupture of the membrane, increasing the solubility and availability of lycopene. Therefore, for better absorption of lycopene in the body, it should be ingested cooked, like in tomato sauce and foods.

1. INTRODUCTION

Lycopene, ((6E, 8E, 10E, 12E, 14E, 16E, 18E, 20E, 22E, 24E, 26E) - 2, 6, 10, 14, 19, 23, 27, 31 – octamethyldotriaconta - 2, 6, 8, 10, 12, 14, 16, 18, 20, 22, 24, 26, 30 - tridecaene) also known as γ, γ - carotene, all-trans lycopene, and (all-E) lycopene is one of the 600 natural pigments called carotenoids, widely distributed in nature [1]. Along with other compounds of this group are found in plants, animals and bacteria, and provide hues ranging from yellow to deep red in some fruits and flowers [2, 3]. Some of these pigments are micronutrients in foods that may exist in the form of different chemical structures and bioactive substances which have beneficial effects on human health. Both the physico-chemical properties, as the functions and actions of carotenoids are directly related to structure type.

Despite having similar chemical formulas, the chemical structures of the carotenoids differ and consequently end up performing some biological activities in the human body [4]. Some carotenoids possess cyclic structures with β-ionone ring, which confers the pro-vitamin A. Lycopene does not present pro-vitamin A activity, it is acyclic, however, it is believed that it is an important antioxidant agent capable to reduce free radicals[6]. Studies have associated this feature the treatment of several diseases, especially in the prevention and treatment of many types of cancer [6].

The lycopene is soluble and hydrophobic, has a system of conjugated double bonds, which gives pink to red color. Watermelon, red guava, papaya, cherry and tomato are the main natural sources of lycopene [7]. In nature the predominant isomeric form of lycopene is trans, to be more stable, however, the cis isomers are most commonly found in plasma and absorbed by the human body, due to its greater micelles solubility [1, 8].

2. Chemical Properties of Lycopene

Lycopene, as such as carotenoids are tetraterpenóides (C40) consist of eight isoprenoid (C5), joined by tail-head joints, except in the center where the union is tail-tail, forming symmetrical molecules. Six carbons separating the central methyl groups, while five separate too. Its main feature is the system long conjugated double bonds which functions as a light-absorbing chromophore responsible for food coloring which is present.

Lycopene has basic linear structure, but other carotenoids cyclic rings may contain one or more extremities, are hydrogenated, dehydrogenated, undergoing chain extension or shortening, isomerization, rearrangements, introduction of oxygen-containing groups, or combinations of these modifications, forming a wide variety of structures [7, 9]. They are divided into two major groups according to their chemical composition: xanthophylls group, which is formed by carotenoids having the presence of at least one oxygen atom bound to the molecule. Some representatives of this group are zeaxanthin, lutein and the α- and β-cryptoxanthin. The other group is the carotenes, which are part of the carotenoids composed solely of carbon and hydrogen, such as β-carotene, α-carotene and lycopene [7, 10].

The lycopene molecule ($C_{40}H_{56}$) is acyclic and features, as well as other tetraterpenoids 40 carbon atoms arranged in a straight chain highly unsaturated, double bonds totaling 13, 11 which are coupled and the other two are at the ends of this molecule (Figure 1) [11, 12]. This arrangement makes it lacks the β-ionone ring, present in approximately 50 of the more than 600 known carotenoids, which would allow its conversion into vitamin A [12-14].

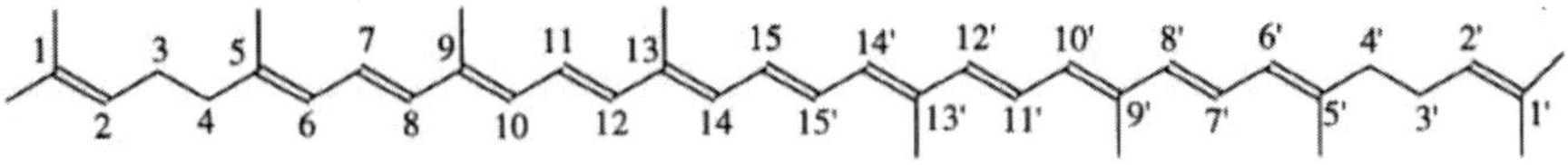

Figure 1. Molecular structure of lycopene.

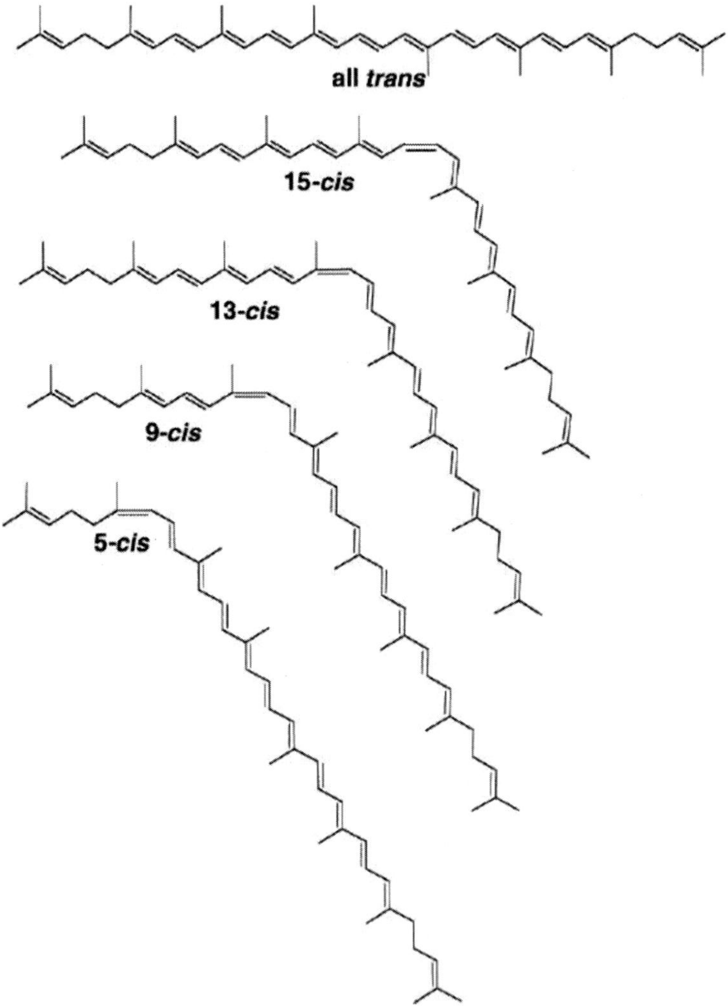

Figure 2. Common isomeric forms of lycopene. Adapted by Agarwal and Rao[5].

Lycopene occurs naturally in all-trans form, due to their greater stability but has a large variety of geometrical isomers, and may take mono-cis or poly-cis after being exposed to high temperatures, catalysts, acids or contact with oxygen. The physical and chemical properties of cis lycopene isomers differ in the all-trans form, as in the cis configuration is more soluble in hydrocarbon solvents [16].

In human plasma there is a match for the concentration of lycopene cis-isomers and trans-lycopene isomers being 50% of the cis form. The most common forms are isomeric all- trans, 9-cis 5-cis 13-cis and 15-cis [17] showed in Figure 2.

One of the main functions of the lycopene in the body is the antioxidant, as such as free radicals and oxygen singlet quenching [18]. Halliwell defined as antioxidant substances found at low concentrations compared with the concentration of an oxidizable substrate in order to cause the delay or prevention of oxidizing said substract [19]. They prevent the sequence of reactions of free radical chain, removing intermediates, and inhibit other oxidation reactions which may damage cells and lead to cancer [20]. To control the amount of the radical are used antioxidants, able to neutralize excess present in the body.

It is believed that lycopene is the main antioxidant in the group of carotenoids, given its control potential of damaging free radicals, especially against the most aggressive reactive oxygen species, the singlet oxygen [21].

3. Natural and Synthetic Sources of Lycopene

Lycopene is found naturally in a short list of fruits and vegetables that have color ranging from pink to deep red. Some of these foods are: tomato (Lycopersicon esculentum), the red guava (Psidium guajava), watermelon (Citrullus lanatus), papaya (Carica papaya) and cherry (Eugenia uniflora). Tomato (in its natural form) is the one with the highest concentration of lycopene which according to the degree of ripening can be increased [6, 17, 22]. Products derived from tomatoes as juice, paste, ketchup, sauce and soup are also considered important sources of obtaining lycopene in the diet [17] (Table 1).

In fruits and vegetables in general, the available amount of lycopene will also depend on many factors such as variety, climatic effect, planting location, post harvest treatment and storage. The more red food coloring is, the higher the levels of lycopene, which is found in higher concentrations than in the bark in relation to the pulp fruit [23].

Table 1. Naturally occurring levels of lycopene in food

Food category	mg/kg
	Natural occurrence
Baked beans	3.2
Grapefruit	12.1
Watermelon	46.8
Meat dishes	28.4
Pasta Dishes	69.1
Pizza	32.6
Soup (other than tomato soups)[1]	12.4
Tomato concentrate	584.3
Tomato ketchup	154.6
Tomato canned	77.4
Tomato juice and vegetables juice cocktail tomato juice-based	90.2
Tomato paste/puree	254.3
Tomato raw	54.9
Tomato soup	62
Tomato sun-dried	407.5
Tomato sauce	153
Red sweet sauces (barbecue sauces)	50
Vegetable dishes	31

[1] tomato only as one of various ingredients, for instance in dry or canned vegetable soups. Fonte: EFSA Journal 2010 [24].

4. Biological Activity of Lycopene

The consumption of fruits and vegetables that have carotenoids is almost universally associated with decreased risk of many diseases such as heart disease and cancer [21, 25]. Many researches and studies claim that lycopene may play a beneficial effect on human health and its possible action in the body contributes to the prevention and treatment of various cancers, especially prostate, lung and stomach. In addition, studies have shown that lycopene can be beneficial in cases of cancer of the esophagus, pancreas, colon and rectum [11, 26]. This relationship is due to the large lycopene antioxidant potential, acting in the control of oxidative stress related to the origin of cancer [27].

After being absorbed, the lycopene accumulates in tissues and body fluids such as blood serum prostate and mainly acting at these sites. The mode of action of lycopene in the prevention of cardiovascular disease is related to the ability to oxidation protect in cells and lipoproteins [15]. Still, relates that the increased consumption of foods that have lycopene and other carotenoids in its composition, such as tomatoes and their derivatives, may represent a decreased risk of developing prostate cancer men, working in the prevention of such pathology [28].

Studies suggest that products derived from tomatoes, especially those who are subjected to thermal processing have a higher antioxidant potential in the body, due to the conversion of trans-lycopene cis-lycopene to isomerism. The thermal process has the ability to increase the nutritional value of tomatoes, the lycopene content increases and antioxidant activity [29].

REFERENCES

[1] Mangels, Ann Reed et al. Carotenoid content of fruits and vegetables: An evaluation of analytic data. *Journal Of The American Dietetic Association.* Chicago, p. 284-296. Mar. 1993. apud Landrum, John T. (Ed.). *Physical, Chemical, and Biological Functions and Properties.* Boca Raton: Crc Press, p.568, 2009.

[2] Vitale, Arturo Alberto; Bernatene, Eduardo Alberto; Pomilio, Alicia Beatriz. Carotenoides en quimioprevención: Licopeno. *Acta Bioquímica Clínica Latinoamericana.* La Plata, v. 44, n. 2, p. 195-238, 2010.

[3] Eldahshan, Omayma; Singab, Abdel Nasser B. Carotenoids. *Journal of Pharmacognosy And Phytochemistry.* Devipura, v. 2, n. 1, p. 225-234, 2013.

[4] Rodriguez-Amaya, Delia B.; Kimura, Mieko. *Harvest Plus Handbook for Carotenoid Analysis.* Washington: International Food Policy Research Institute, p.58, 2004.

[5] Agarwal, Sanjiv; RAO, Akkinappally Venketeshwer. Tomato lycopene and its role in human health and chronic diseases. *Canadian Medical Association Journal*, v. 163, n. 6, p.739-744, 2000.

[6] Palozza, Paola et al. Tomato Lycopene and Lung Cancer Prevention: From Experimental to Human Studies. *Cancers*, v. 3, n. 2, 2333-2357, 2011.

[7] Rodriguez-Amaya, Delia B. *A Guide to Carotenoid Analysis in Foods.* Washington: Inn Press, p.64, 2001.

[8] Boileau, Amy C. et al. Cis-Lycopene Is More Bioavailable Than Trans-Lycopene In Vitro and In Vivo in Lymph-Cannulated Ferrets. *The Journal of Nutrition*, v. 129, n. 6, 1176-1181, 1999.

[9] Rao, A. Venket.; RAO, Leticia G. Carotenoids and Human Health. *Pharmacological Research*, v. 55, 207-216, 2007.

[10] Stahl, Wilhelm; SIES, Helmut. Bioactivity and protective effects of natural carotenoids. *Biochimica Et Biophysica Acta - Molecular Basis Of Disease*, v. 1740, n. 2, 101-107, 2005.

[11] Rao, Leticia G; GUNS, Emma; RAO, A. Venket. Lycopene: Its role in human health and disease. *Agrofood Industry Hi-tech*, p.25-30, 2003.

[12] Shi, John et al. Stability and Synergistic Effect of Antioxidative Properties of Lycopene and Other Active Components. *Critical Reviews In Food Science And Nutrition*, v. 44, n. 1, 559-573, 2004.

[13] Rodriguez-Amaya, Delia B. *Carotenoids and Food Preparation: The Retention of Provitamin A Carotenoids in Prepared, Processed, and Stored Foods.* Boston: John Snow Incorporated/omni Project, p.93, 1997

[14] Khachik, Frederick et al. Transformations of Selected Carotenoids in Plasma, Liver, and Ocular Tissues of Humans and in Nonprimate Animal Models. *Investigative Ophtalmology And Visual Science*, v. 43, n. 11, 3383-3392, 2002.

[15] Bramley, Peter. Is lycopene beneficial to human health? *Phytochemistry*, v. 54, n. 3, 233-236, 2000.

[16] Shi, John; Mazza, Giuseppe; Maguer, Marc Le. *Functional Foods: Biochemical and Processing Aspects.* Boca Raton: Crc Press, p. 432, 2002.

[17] Rao, A. Venket; Agarwal, Sanjiv. Role of Antioxidant Lycopene in Cancer and Heart Disease. *Journal Of The American College Of Nutrition,* v. 19, n. 5, 563-569, 2000.

[18] Sies, Helmut; Stahl, Wilhelm. Vitamins E and C, a-carotene, and other carotenoids as antioxidants1-3. *American Journal Of Clinical Nutrition,* v. 62, n. 6, 1315-1321, 1995.

[19] Halliwell, Barry. *Free Radicals and other reactive species in Disease.* Singapore: Encyclopedia Of Life Sciences, p.7, 2001.

[20] Hamid, A. A. et al. Antioxidants: Its medicinal and pharmacological applications. *African Journal Of Pure And Applied Chemistry*, 142-141, 2010.

[21] Erdman, Jr, John, W; Ford, Nikki, A; Lindshield, Brian L. Are the health attributes of lycopene related to its antioxidant function? *Archives Of Biochemistry And Biophysics*, v. 483, n. 2, 229-235, 2009.

[22] Ševčíková S., Skřivan M., Dlouhá G. The effect of lycopene supplementation on lipid profile and meat quality of broiler chickens. *Czech Journal Of Animal Science*, v. 53, n. 10, 431-440, 2008.
[23] Preedy, Victor R; Watson, Ronald R. (Ed.). *Tomatoes and Tomato Products: Nutritional, Medicinal and Therapeutic Properties.* Enfield: Science Publishers, p. 665, 2008.
[24] Italy. European Food Safety Authority. . *Revised exposure assessment for lycopene as a food colour.* Parma: Efsa Journal, p.26, 2010.
[25] Hoppe, Peter P. et al. Synthetic and tomato-based lycopene have identical bioavailability in humans. *European Journal Of Nutrition.* 272-278, 2003.
[26] Giovannucci, Edward. Tomatoes, Tomato-Based Products, Lycopene, and Cancer: Review of the Epidemiologic Literature. *Journal Of National Cancer Institute*, v. 91, n. 4, 317-331, 1999.
[27] Devaraj, Sridevi et al. A Dose-Response Study on the Effects of Purified Lycopene Supplementation on Biomarkers of Oxidative Stress. *The Journal Of The American College Of Nutrition*, 267-273, 2008.
[28] Giovannucci, Edward et al. Intake of Carotenoids and Retino in Relation to Risk of Prostate Cancer. *Journal Of The National Cancer Institute*, 1767-1776, 1995.
[29] Dewanto, Veronica et al. Thermal Processing Enhances the Nutritional Value of Tomatoes by Increasing Total Antioxidant Activity. *Journal Of Agricultural And Food Chemistry*, v. 50, n. 10, 3010-3014, 2002.

In: Lycopene
Editor: Jacob R. Bailey

ISBN: 978-1-63117-927-3

Chapter 5

USE OF ENCAPSULATION TECHNOLOGY FOR IMPROVING THE STABILITY OF LYCOPENE

***Zhen-Xing Tang*[1,*], *Cai-E Wu*[1] *and Lu-E Shi*[2]**
[1]College of Light Industry Science and Engineering,
Nanjing Forestry University, Nanjing, Jiangsu, China
[2]College of Life and Environmental Sciences,
Hangzhou Normal University, Hangzhou, Zhejiang, China

ABSTRACT

Lycopene is a natural red pigment mainly found in tomato, tomato products, and lycopene-carrots, which belongs to the group of carotenoids. It shows a very high quenching rate of reactive singlet oxygen, due to its apolar and acyclic molecular structure. Epidemiological evidence has suggested that lycopene can protect individuals from colorectal cancer and prostate cancer. However, as many members among the group of carotenoids, lycopene is very sensitive to heat, oxidation and light, due to their unsaturated chemical structures. It will limit the activity and potential health benefits of lycopene. Therefore, in order to improve the stability of lycopene, the processes for improving the stability of lycopene, must be developed and investigated, which will be mainly discussed in this chapter.

* Corresponding author: Dr. Zhen-Xing Tang, E-mail: tangzhenxing@126.com

INTRODUCTION

Lycopene is a natural carotenoid that imparts red color to various fruits and vegetables such as tomato, rosehip, watermelon and pink grape fruit (Holden et al., 2003; Ribeiro et al., 2010; Srivastava and Srivastava, 2013). Tomato, especially deep red fresh tomato, is considered as the most important source of lycopene. Due to the anti-oxidative properties of lycopene, many studies regarding its protective effects against cardiovascular diseases and certain types of cancers, as well as other degenerative illnesses, have been carried out in last few years (Gerster, 1997; Giovanucci et al., 1995; Grant, 1999; Moselhi and Al Mslmani, 2008; Schwarz et al., 2008; Sesso et al., 2003, 2004; Slattery et al., 2000; Srivastava and Srivastava, 2013; Vrieling et al., 2007). Lycopene can inhibit the destructive effect of reactive oxygen species. It has been regarded as the most effective singlet oxygen quencher among all carotenoids due to its apolar and acyclic molecular structure. Lycopene participates in many chemical reactions which prevent the degradation of critical cellular biomolecules, including lipids, proteins and DNA (Di Mascio et al., 1989; Srivastava and Srivastava, 2013). *In vitro* studies have also shown that lycopene can stimulate bone formation and mineralization. The results may be related to prevention of osteoporosis (Kim et al., 2003; Yamaguchi and Uchiyama, 2003; 2004; Yamaguchi et al., 2005). After consumed, lycopene is deposited in the lungs, liver, prostate gland, colon and skin in the human body. Its concentration in body tissues tend to be higher than that of all other carotenoids (Ferreira et al., 2000; Rao et al., 1998; Shi and Le Maguer, 2000; Srivastava and Srivastava, 2013).

The increasing awareness of the health benefits of lycopene results in its increased uses as a neutraceutical and in functional foods. However, lycopene is very sensitive to heat, oxidation, and light due to their unsaturated chemical structure, and easily degraded via isomerization or auto-oxidation that affect its stability (Garret et al., 1999; Giovanucci et al., 1995; Nguyen and Schwartz, 1998; Ribeiro et al., 2010; Shi, 2000). Using encapsulation technology, the stability of lycopene can be considerably improved (Ax et al., 2003; Engel et al., 2005; Horn and Rieger, 2001; McClements et al., 2007; Ribeiro et al., 2003, 2005 a, b, 2008, 2010). In order to achieve optimized the stability of lycopene, encapsulation technology has been developed and investigated, which will be discussed in this chapter.

METHODS OF ENCAPSULATING LYCOPENE

Since lycopene is very expensive, in order to reduce the cost of the final products, the cost-effective processes are desirable to produce lycopene formulations. Optimization of encapsulation conditions may prevent lycopene oxidation. The most important variables to achieve droplets size and desired color for lycopene application are choice of optimal process, emulsifier(s), and other ingredients (Lüddecke et al., 2004; Ribeiro et al., 2010).

OIL-IN-WATER (O/W) EMULSION

Lycopene emulsions can be produced using various emulsification processes. For preparing lycopene-loaded O/W emulsion, lycopene is dissolved in a vegetable oil or in an apolar solvent, and subsequently emulsified with an aqueous phase containing an emulsifier to stabilize the droplets (Ribeiro et al., 2010). Lycopene O/W emulsions have been prepared by high-pressure homogenization, and phase inversion temperature (PIT) method.

High-Pressure Homogenization

High-pressure homogenization is an ideal mechanical method for producing droplet sizes below 1 μm. It consists of a high-pressure pump and a homogenizing nozzle. The pump creates the pressure, which is then transferred within the nozzle to kinetic energy being responsible for droplet disintegration. The lycopene molecules themselves do not have any relevant interfacial activity and do not change system viscosity. Thus droplet disruption is not influenced by dissolved lycopene molecules (Ax et al., 2003; Ribeiro et al., 2010; Velikov and Pelan, 2008). Droplet size distribution depends on applied emulsifiers, oil type, homogenizing nozzle geometry, and homogenization pressure. Ax et al., (2003) studied the stability of lycopene in O/W emulsions prepared by high pressure homogenization. The results showed that it was sensitive to oxidative conditions and easily underwent isomerization at high temperatures. Lycopene degradation followed first-order kinetics. Higher temperatures were directly corrected with increasing lycopene losses.

Kim et al., (2014) prepared lycopene nanoemulsions using homogenization evaporation technique and optimized lycopene nanoemulsion process. Lycopene nanoemulsion with the highest stability and smallest size was obtained after 3 homogenization cycle with 0.085 mg/mL of lycopene extract and 0.7 mg/mL of Tween 20. A degradation study of lycopene in the beverage showed that nanoemulsions entrapping lycopene are stable at both 4 and 20 °C. This study provided important information about the preparation for lycopene nanoemulsions and their beverage applications.

Guo et al., (2014) presented a new way to prepare lycopene microcapsules with ultrasonic emulsification method. The results suggested that the microcapsules had a regular rounded shape and lycopene was embedded in the wall system. Mannitol was showed excellent efficiency on retaining the stability of lycopene microcapsules. The results would be helpful to promote the application of lycopene.

Phase Inversion Temperature (PIT) Method

O/W emulsions containing ethoxylated non-ionic emulsifiers may undergo a phase inversion to W/O emulsions upon heating. This technique is a thermodynamic phenomenon that produces nano-dispersed transparent O/W emulsions with narrow droplet size distributions (Engels et al., 1995; Ribeiro et al., 2010). Ribeiro and Schubert (2004) produced nanosized carotenoid-loaded O/W emulsions using cetostearyl poly(oxyethylene glycol) and glyceryl monostearate as the emulsifying agents by PIT method. The PIT emulsion was prepared by heating up to 74 °C and subsequently rapid cooling down to 25 °C. The total lycopene concentration was around 3.5 g/L emulsion. No detectable degradation of lycopene was observed after a month of storage. The mean diameter of the finely dispersed droplets was around 80 nm. These data suggest that PIT emulsions could be an excellent vehicle for delivering lycopene.

Production of Lycopene Nanoparticles by Precipitation

Three different methods can be applied to produce lycopene-loaded organic nanoparticles: emulsification evaporation, emulsification diffusion and solvent displacement methods. Emulsification evaporation technique is applied when an apolar organic solvent is used. Solvent-displacement method is appropriated when a polar solvent (acetone, ethanol) is used as the dispersed phase. The nanodroplets formation process is related to the decreasing droplet size due to the rapid diffusion of solvent to the aqueous phase. Emulsification diffusion method, which uses amphiphilic solvent (benzyl alcohol, ethyl acetate) as the dispersed phase, can also be successfully applied to produce lycopene nanoemulsions with remarkable narrow droplet size distributions and small droplets size (Ribeiro et al., 2010). Bortlik et al., (2005) patented a highly bioavailable whey protein-based formulation which may be applied to encapsulate lycopene. Polar organic solvent were used for dissolving the active molecule. When lycopene was encapsulated into the whey protein matrix, it exhibited similar bioavailability in plasma and buccal mucosa cells in humans compared to tomato paste (Richelle et al., 2002).

Lycopene Nanoparticles Produced by Supercritical Fluid

Rapid expansion of supercritical solution into aqueous solution (RESSAS) is a novel process for producing nanoparticles in aqueous solution. Stable lycopene nanoparticles can be produced by RESSAS. In this method, a mixture of a poorly water soluble substance and a supercritical fluid, such as carbon dioxide (CO_2), or nitrous oxide (N_2O), is expanded into an aqueous solution containing a surfactant, which may impede growth and/or agglomeration of nanoparticles which can happen due to collisions in the capillary nozzle into the expansion chamber (Türk and Lietzow, 2004; Ribeiro et al., 2010). Miguel et al., (2006) used the supercritical fluid process to obtain lycopene needle-like particles with the size between 10 μm and 80 μm. They found that particle size increased with an increase in pressure and rate ratio (anti-solvent to solution flow), but decreased with increased in temperature and initial concentration.

Lycopene Encapsulation Using Cyclodextrins

Cyclodextrins are able to serve as host molecules to encapsulate lycopene (Hedges et al., 1995; Matioli and Rodriguez-Amaya, 2003; Ribeiro et al.,

2010; Shahidi and Han, 1993). Blanch et al., (2007) used supercritical fluid extraction and a conventional method for entrapping lycopene into cyclodextrin molecules. Higher encapsulation yield (94 %) from the conventional method was obtained compared to that from supercritical fluid extraction method. However, the supercritical fluid approach offered numerous advantages such as the possibility of performing the extraction, fractionation and encapsulation of lycopene from tomato in one step, shortening notably the overall procedure time and minimizing the sample handling. The obtained β-Cyclodextrin–lycopene complex was chemically and physically stable over 6 months. The results showed that cyclodextrins can be used to increase lycopene solubility and improve its absorption, protect against light and oxygen, and may prevent off-flavor.

PREPARATION OF MICROCAPSULES BY DRYING PROCESSES

Drying can also be considered as an important unit operation to transform moist or fluid lycopene formulations into dry solids. It allows the production of lycopene microcapsules with a certain structure, size, and shape (Ribeiro et al., 2010).

Chiu et al., (2007) reported an interesting encapsulation study done with 4.8 % lycopene extract from tomato pulp waste. 4.5 % Gelatin and 10 % poly (g-glutamic acid) (g-PGA) were used as carrier matrix. The results showed that microcapsules around 39 mm were obtained, and freeze-drying caused a loss of 23.5 % lycopene.

Silva et al., (2012) prepared lycopene microcapsules using gelatin and pectin as wall materials. The results of this work indicated that bovine gelatin and pectin had the potential for use as wall materials in the complex coacervation process. Spherical microcapsules containing lycopene were obtained. However, encapsulation did not attenuate lycopene loss during storage, probably because the microcapsules lose their initial structure or crumble during the freezing and/or the lyophilization procedure.

Montenegro et al., (2007) spray-dried lycopene-loaded O/W emulsion stabilized with a gum arabic and sucrose mixture. Vitamins degradation has slowed down by 45 % in presence of lycopene formulation.

Shu et al., (2006) developed a spray drying method for preparation of lycopene microcapsules with a combination of gelatin and sucrose as wall materials. Encapsulation yield and efficiency were significantly affected by the ratio of core and wall materials, the ratio of gelatin and sucrose, homogenization pressure, temperatures during drying process, feed temperature and lycopene purity. The optimal condition was determined as follows: the ratio of gelatin/sucrose of 3/7 and the ratio of core and wall material of 1/4, feed temperature of 55 ℃, inlet temperature of 190 ℃, homogenization pressure of 40 MPa. The highest values of encapsulation yield and efficiency found were about 91 % and 82 %, respectively.

Rocha et al., (2012) encapsulated lycopene by spray drying, using a modified starch as an encapsulating agent. Encapsulation efficiency values varied between 21 and 29 %. The microcapsules had a rounded outer surface with the formation of concavities. The stability test revealed that encapsulation offered greater protection to lycopene compared to its free form and it was observed that microcapsules were able to release lycopenes in the controlled rate.

Xue et al., (2013) used zein as coating material to encapsulate the tomato oleoresin by spray-drying. Moisture content, particle yield and encapsulation efficiency of powder increased with an increase in zein concentration. Microcapsules prepared by spray-drying had poor flowing properties. The controlled release results showed that zein particles could protect most of lycopene from being released in the stomach.

THE STABILITY OF LYCOPENE IN FOOD FORMULATIONS

Experiments investigating the stability of lycopene emulsions have been carried out. Droplet size and its distribution are important characteristics of lycopene dispersions, which are associated with many properties, such as shelf-life stability, homogeneity, creaminess, viscosity, and color (Ribeiro et al., 2010). Lycopene-based products can provide health benefits by improving antioxidant levels in individuals, so-called “functional foods”, are, for example, beverages, breads, candies, cookies, dairy products (ice-cream, cheddar cheese, etc.), dry cake mix, salad dressings, macaroni products, mayonnaise, savory, shortenings, soups, and spreads (Leuenberger, 2007; Ribeiro et al., 2010; Santipanichwong and Suphantharika, 2007; Upritchard et al., 2003). When lycopene emulsion-based formulations are mixed with products as foods, they have to remain stable. Ribeiro et al., (2003) studied the stability of the lycopene formulations in three different products: skimmed milk, orange juice, and water (control). It was found that lycopene stability strongly depended on the food system. In orange juice (pH < 4.5), the lycopene was particularly stable, whereas in milk (pH ≅ 6.6) and water (pH 7), respectively. Stability of lycopene did not depend on the emulsifier type. Antioxidants can be used to protect O/W emulsions against degradation over a long period of time. Addition of α-tocopherol to the dispersed oil phase, allowed for prolonged lycopene stability in all three food systems. Coalescence of oil droplets was not observed in any of the systems investigated over 3 weeks. O/W emulsions containing lycopene and vitamin C, and stabilized by Tween 20 was also applied in full- and low-fat yogurt, 3.8 % and 1.5 % fat, respectively. It remained stable for 3 weeks in the full-fat yogurt, however, approximately 20 % of lycopene degradation could be observed in the low fat one over the same storage time.

Ribeiro et al., (2006) have found decreased lycopene degradation in O/W emulsions stabilized with a combination of emulsifiers, especially for a combination of a whey protein isolate and a sugar ester, compared to emulsions prepared with a single emulsifier.

Physical stability of lycopene emulsions includes stability against droplet coalescence and flocculation or aggregation. Droplets can be stabilized against these by emulsifier molecules adsorbed at the interfaces and/or by electrical charges. The droplets zeta-potential is a measure of stabilizing electrostatic interaction forces between the droplets. Ribeiro et al., (2010) measured the droplet zeta-potential of O/W emulsions containing lycopene, prepared with combination of whey protein isolates, with either Tween 20 or sucrose laurate. At the lowest pH the droplets had a relatively high positive zeta-potential because pH was below the isoelectric point of whey proteins. When pH was increased, the magnitude of the positive charge on the droplets decreased. Eventually the zeta-potential of the droplets became zero (isoelectric point at a pH between 4 and 5). A further increase in pH caused the droplets to gain a net negative charge. Droplet agglomeration could be observed after 24 h in samples at pH 4. Droplet flocculation/aggregation caused a significant increase in the creaming of the droplets in the emulsions.

STABILITY OF LYCOPENE FORMULATIONS DURING PASSAGE

The *in vivo* efficacy of formulations containing functional ingredients depends on the one hand, on the stability of the emulsions during processing and storage, and on the other hand, on their behavior while passing through the gastro-intestinal tract. After oral ingestion of the emulsions, the chemical and enzymatic conditions in the stomach and small intestine can lead to extensive chemical and physical changes in the emulsion system and thereby influence the absorption of the active substance at the area of resorption (Ribeiro et al., 2010).

Ax et al., (2003) studied the effect of digestion on the stability of lycopene emulsion stabilized by Tween 20. As an indicator of the stability of the emulsion against aggregation or coalescence, particle size distributions and zeta potential were measured. The zeta potential changed from −35 mV at the beginning to about −20 mV after stomach incubation. No significant change in droplet size distribution after processing in the stomach was observed. Under the conditions in the small intestine, addition of pancreatic lipase to the emulsion led, as a result of lipolysis, to a prompt decline in pH. The appearance of the emulsions changed from opaque to translucent, they darkened in color, and viscosity changes occurred. All these observations indicated that the dispersed oil droplets were destroyed.

Conclusion

This chapter has shown many methods to encapsulate lycopene. Optimal powder or liquid lycopene formulations allow the production of the stability after their preparation, storage, and application in food systems. These requirements are very important for delivering the lycopene into the gastro-intestinal tract, and later for being absorbed. The ideal encapsulation process depends on specific application, market requirement, and current regulations. However, most of these encapsulation methods are only promising in a laboratory scale.

Encapsulation still has to face many challenges for its application on an industrial scale. Even lycopene encapsulation faces many challenges, it is evident that lycopene market has a strong future as the consumers demand is increasing. Good hopes are also visualised for the encapsulation of lycopene in the future.

REFERENCES

Ax, K., Mayer-Miebach, E., Link, B., Schuchmann, H. and Schubert, H. (2003) Stability of lycopene in oil-in-water emulsions. *Eng. Life Sci. 3,* 199–201.

Blanch, G. P., Castillo, M. L. R., Caja, M. M., Pérez-Méndez, M. and Sánchez-Cortés, S. (2007) Stabilization of all-trans-lycopene from tomato by encapsulation using cyclodextrins. *Food Chem.* 105, 1335–1341.

Bortlik, K., Saucy, F., Duruz, E., Richelle, M., Lambelet, P., Baur. M. and Pfeifer, A. M. A. (2005) Primary composition comprising a lipophilic bioactive compound. *US* 2005/0106219 A1.

Chiu, Y. T., Chiu, C. P., Chien, J. T., Ho, G. H., Yang, J. and Chen, B. H. (2007) Encapsulation of lycopene extract from tomato pulp waste with gelatin and poly(g-glutamic acid) as carrier. *J. Agric. Food Chem.* 55, 5123–5130.

Di Mascio, P., Kaiser, S. and Sies, H. (1989) Lycopene as the most efficient biological carotenoid singlet oxygen quencher. *Arch. Biochem. Biophys.* 274, 532–538.

Engels, T., Förster, T. and von Rybinski, W. (1995) The influence of coemulsifier type on the stability of oil-in-water emulsions. *Colloids Surf* A 99,141–149.

Engel, R., Ribeiro, H. S., Briviba, K., Rechkemmer, G., Schuchmann, H. P. and Schubert, H. (2005) Untersuchungen zur Formulierung von Carotinoiden und Phytosterolen in Mizellen zur Verbesserung der ernährungsphysiologischen Wirksamkeit. *German Nutr. Soc.* 7, 82.

Ferreira, A. L. A., Yeum, K. J., Liu, C., Smith, D., Krinsky, N. I., Wang, X. D. and Russel, R. M. (2000) Tissue distribution of lycopene in ferrets and rats after lycopene supplementation. *J. Nutr.* 130, 1256–1260.

Garret, D. A., Failla, M. L. and Sarama, R. J. (1999) Development of an in vitro digestion method to assess carotenoid bioavailability from meals. *J. Agric. Food Chem.* 47, 4301–4309.

Gerster, H. (1997) The potential role of lycopene for human health. *J. Am. Coll. Nutr.* 16, 109–126.

Giovanucci, E., Ascherio, A., Rimm, E. B., Stampfer, M. M., Colditz, G. A. and Willett, W. C. (1995) Intake of carotenoids and retinal in relationship to risk of prostate cancer. *J. Natl. Cancer Inst.* 87, 1767–1776.

Grant, W. B. (1999) An ecologic study of dietary links to prostate cancer. *Altern Med. Rev.* 4, 167–169.

Guo, H., Huang, Y., Qian, J. Q., Gong, Q. Y. and Tang, Y. (2014) Optimization of technological parameters for preparation of lycopene microcapsules. *J. Food Sci. Technol.* 51, 1318-1325.

Hedges, A. R., Shieh, W. J. and Sikorski, T. (1995) Use of cyclodextrin for encapsulation in the use and treatment of food products. In: Risch, S. J. and Reineccius, G. A. (eds) Encapsulation and controlled release of food ingredients. ACS Symposium Series 590, USA.

Horn, D. and Rieger, J. (2001) Organische Nanopartikel in wässriger Phase – Theorie, Experiment und Anwendung. *Angewandte Chemie* 113, 4460–4492.

Kawakatsu, T., Kikuchi, Y. and Nakajima, M. (1997) Regular-sized cell creation in microchannel emulsification by visual microprocessing method. *J. Am. Oil Chem. Soc.* 74, 317–321.

Kikuchi, Y., Kikuchi, H. E., Kuboi, Y. and Nakajima, M. (2000) Micro- and nanotechnology for biomedical and environmental applications. *Proc. SPIE* 3912, 95–104.

Kim, L., Rao, A. V. and Rao L. G. (2003) Lycopene II – Effect on osteoblasts: The carotenoid lycopene stimulates cell proliferation and alkaline phosphatise activity of SaOS-2 cells. *J. Med. Food* 6, 79–86.

Kim, S. O., Ha, T. V. A., Choi, Y. J. and Ko, S. (2014) Optimization of Homogenization-evaporation process for lycopene nanoemulsion production and its beverage applications. *J. Food Sci.* 79, N1604-N1610.

Leuenberger, B. H. (2007) Conceptual design of carotenoid product forms. In: Bröckel, U., Meier, W. and Wagner, G. (eds) Product design and engineering, Vol. 2: Rawmaterials, additives and applications. Wiley-VCH Verlag, Weinheim, pp 539–567.

Lüddecke, E., Auweter, H. and Schweikert, L. (2004) Use of carotenoid aggregates as colorants. *US* 6,827,941 B1.

Matioli, G. and Rodriguez-Amaya, D. B. (2003) Microencapsulation of lycopene with cyclodextrins. *Ciência e Technologia de Alimentos* 23,102–105.

McClements, D. J., Decker, E. A. and Weiss, J. (2007) Emulsion-based delivery systems for lipophilic bioactive components. *J. Food Sci.* 72, R109–R124.

Miguel, F., Martín, A., Gamse, T. and Cocero, M. J. (2006) Supercritical anti solvent precipitation of lycopene: effect of the operating parameters. *J. Supercrit Fluids* 36, 225–235.

Montenegro, M. A., Nunes, I. L., Mercadante, A. Z. and Borsarelli, C. D. (2007) Photoprotection of vitamins in skimmed milk by an aqueous soluble lycopene-gum arabic microcapsule. *J. Agric. Food Chem.* 55, 323–329.

Moselhi, S. S. and Al Mslmani, M. A. B. (2008) Chemopreventive effect of lycopene alone or with melatonin against the genesis of oxidative stress and mammary tumors induced by 7, 12 dimethyl (a) benz-anthracene in sprague dawely female rats. *Mol. Cell Biochem.* 319, 175–180.

Nguyen, M. L. and Schwartz, S. J. (1998) Lycopene stability during food processing. *Exp. Biol. Med.* 218,101–105.

Rao, A. V., Waseem, Z. and Agarwal, S. (1998) Lycopene content of tomato and tomato products and their contribution to dietary lycopene. *Food Res. Int* 31,737–741.

Ribeiro, H. S., Ax, K. and Schubert, H. (2003) Stability of lycopene emulsions in food systems. *J. Food Sci.* 68, 2730–2734.

Ribeiro, H. S., Chu, B. S., Ichikawa, S. and Nakajima, M. (2008) Preparation of nanoemulsions containing β-carotene. *Food Hydrocolloid* 22, 12–17.

Ribeiro, H. S., Cruz, R. C. D. and Schubert, H. (2005a) Biliquid foams containing carotenoids. *Eng. Life Sci.* 5, 84–88.

Ribeiro, H. S., Guerrero, J. M. M., Briviba, K., Rechkemmer, G., Schuchmann. H. and Schubert, H. (2006) Cellular uptake of carotenoid-loaded O/W emulsions in colon carcinoma cells in vitro. *J. Agric. Food Chem.* 54, 9366–9369.

Ribeiro, H. S., Rico, L. G., Badolato, G. G. and Schubert, H. (2005b) Production of O/W emulsions containing astaxanthin by repeated premix membrane emulsification. *J. Food Sci.* 70, E117–E123.

Ribeiro, H. S. and Schubert, H. (2004) Lycopene enriched PIT emulsion. In: 3rd Euro fed lipid congress, Scotland.

Ribeiro, H. S., Schuchmann, H. P., Engel, R., Walz, E. and Briviba, K. (2010) Encapsulation of carotenoids. In Zuidam, N.J. and Nedović, V. A. (eds.), Encapsulation Technologies for Active Food Ingredients and Food Processing, Springer Science publication, pp 211-252.

Richelle, M., Bortlik, K., Liardet, S., Hager, C., Lambelet, P., Baur, M., Applegate, L. A. and Offord, E. A. A. (2002) Food-based formulation provides lycopene with the same bioavailability to humans as that from tomato paste. *J. Nutr.* 132, 404–408.

Rocha, G. A., Favaro-Trindade, C. S. and Grosso, C. R. F. (2012) Microencapsulation of lycopene by spray drying: characterization, stability and application of microcapsules. *Food Bioprod. Processing* 90, 37-42.

Santipanichwong, R. and Suphantharika, M. (2007) Carotenoids as colorants in reduced-fat mayonnaise containing spent brewer's yeast β-glucan as a fat replacer. *Food Hydrocolloid* 21,565–574.

Schwarz, S., Obermüller-Jevic, U. C., Hellmis, E., Koch, W., Jacobi, G. and Biesalski, H. K. (2008) Lycopene inhibits disease progression in patients with benign prostate hyperplasia. *J. Nutr.* 138, 49–53.

Sesso, H. D., Buring, J. E., Norkus, E. P. and Gaziano, J. M. (2004) Plasma lycopene, other carotenoids, and retinol and the risk of cardiovascular disease in women. *Am. J. Clin. Nutr.* 79, 47–53.

Sesso, H. D., Liu, S., Gaziano, J. M. and Buring, J. E. (2003) Dietary lycopene, tomato-based food products and cardiovascular disease in women. *J. Nutr.* 133, 2336–234.

Shahidi, F. and Han, X. Q. (1993) Encapsulation of food ingredients. *Crit. Rev. Food Sci. Nutr.* 33, 501–547.

Shi, J. (2000) Lycopene in tomatoes: chemical and physical properties affected by food processing. *Crit. Rev. Food Sci. Nutr.* 40, 1–42.

Shi, J. and Le Maguer, M. (2000) Lycopene in tomatoes: chemical and physical properties affected by food processing. *Crit. Rev. Food Sci. Nutr.* 40, 1–42.

Shu, B., Yu, W., Zhao, Y. and Liu, X. (2006) Study on microencapsulation of lycopene by spray-drying. *J. Food Eng.* 76, 664–669.

Silva, D. F., Favaro-Trindade, C. S., Rocha, G. A. and Thomazini, M. (2012) Microencapsulation of lycopene by gelatin-pectin complex coacervation. *J. Food Process Pres* 36, 185-190.

Slattery, M. L., Benson, J., Curtin, K., Ma, K. N., Schaeffer, D. and Potter, J. D. (2000) Carotenoids and cancer. *Am. J. Clin. Nutr. 71*(2), 575–582.

Srivastava, S. and Srivastava, A. K. (2013) Lycopene: chemistry, biosynthesis, metabolism and degradation under various abiotic parameters. *J. Food Sci. Technol.* DOI 10.1007/s13197-012-0918-2

Sugiura, S., Nakajima M., Iwamoto, S. and Seki, M. (2001) Interfacial tension driven monodispersed droplet formation from microfabricated channel array. *Langmuir* 17, 5562–5566.

Türk, M. and Lietzow, R. (2004). Stabilized nanoparticles of phytosterol by rapid expansion from supercritical solution into aqueous solution. *AAPS PharmSciTech* 5, 1–10.

Upritchard, J. E., Schuurman, C. R. W. C., Wiersma, A., Tijburg, L. B. M., Coolen, S. A. J., Rijken, P. J. and Wiseman, S. A. (2003) Spread supplemented with moderate doses of vitamin E and carotenoids reduces lipid peroxidation in healthy, nonsmoking adults. *Am. J. Clin. Nutr.* 78, 985–992.

Velikov, K. and Pelan, E. (2008) Colloidal delivery systems for micronutrients and nutraceuticals. *Soft Matter* 4, 1964–1980.

Vrieling, A., Voskuil, D. W., Bonfrer, J. M., Korse, C. M., van Doorn, J., Cats, A., Depla, A. C., Timmer, R., Witteman, B. J., van Leeuwen, F. E., van't Veer, L. J., Rookus, M. A. and Kampman, E. (2007) Lycopene supplementation elevates circulating insulin-like growth factor-binding protein-1 and -2 concentrations in persons at greater risk of colorectal cancer. *Am. J. Clin. Nutr.* 86, 1456–1462.

Winterhalter, M. and Lasic, D. D. (1993) Liposome stability and formation: Experimental parameters and theories on the size distribution. *Chem. Phys.* Lipids 64, 35–43.

Xue, F., Li, C., Liu, Y. L., Zhu, X. W., Pan, S. Y. and Wang, L. F. (2013). Encapsulation of tomato oleoresin with zein prepared from corn gluten meal. *J. Food Eng.* 119, 439-445.

Yamaguchi, M., Igarashi, A., Morita, S., Sumida, T. and Sugawara, K. (2005) Relationship between serum β-cryptoxanthin and circulating bone metabolic markers in healthy individuals with the intake of juice (Citrus unshiu) containing β-cryptoxanthin. *J. Health Sci.* 51, 738–743.

Yamaguchi, M. and Uchiyama, S. (2003) Effect of carotenoid on calcium content and alkaline phosphatase activity in rat femoral tissues in vitro: The unique anabolic effect of β-cryptoxanthin. *Biol. Pharm Bull* 43, 281–296.

Yamaguchi, M. and Uchiyama, S. (2004) β-Cryptoxanthin stimulates bone formation and inhibits bone resorption in tissue culture in vitro. *Mol. Cell Biochem.* 258, 137–144.

INDEX

A

B

C

D

E

F

G

H

I

J

K

L

M

N

O

P

Q

R

S

T

U

V

W

Y

Z

β